GW01048588

THE WALTHAM BOOK OF

DOG & CAT NUTRITION

SECOND EDITION

Other Pergamon Publications of Interest

Books

ANDERSON
Nutrition of the Dog and Cat

LANE
Jones's Animal Nursing, 4th Edition

ROBINSON
Genetics for Cat Breeders, 2nd Edition

ROBINSON
Genetics for Dog Breeders

STEPHEN
Trypanosomiasis: a Veterinary Perspective

Review Journals*

Nutrition Research

Progress in Food and Nutrition Science

Research Journals*

Comparative Immunology, Microbiology and Infectious Diseases

International Journal of Parasitology

**Free specimen copies sent on request*

THE WALTHAM BOOK OF

DOG & CAT NUTRITION

SECOND EDITION

A Handbook for Veterinarians and Students

Editor: A. T. B. EDNEY

Formerly of the
Waltham Centre for Pet Nutrition
Waltham-on-the-Wolds, Melton Mowbray
Leicestershire, UK

WALTHAM®
EUROPE'S LEADING
AUTHORITY ON PET NUTRITION

PERGAMON PRESS

OXFORD · NEW YORK · BEIJING · FRANKFURT
SÃO PAULO · SYDNEY · TOKYO · TORONTO

U.K.	Pergamon Press plc, Headington Hill Hall, Oxford OX3 0BW, England
U.S.A.	Pergamon Press, Inc., Maxwell House, Fairview Park, Elmsford, New York 10523, U.S.A.
PEOPLE'S REPUBLIC OF CHINA	Pergamon Press, Room 4037, Qianmen Hotel, Beijing, People's Republic of China
FEDERAL REPUBLIC OF GERMANY	Pergamon Press GmbH, Hammerweg 6, D-6242 Kronberg, Federal Republic of Germany
BRAZIL	Pergamon Editora Ltda, Rua Eça de Queiros, 346, CEP 04011, Paraiso, São Paulo, Brazil
AUSTRALIA	Pergamon Press Australia Pty Ltd., P.O. Box 544, Potts Point, N.S.W. 2011, Australia
JAPAN	Pergamon Press, 5th Floor, Matsuoka Central Building, 1-7-1 Nishishinjuku, Shinjuku-ku, Tokyo 160, Japan
CANADA	Pergamon Press Canada Ltd., Suite No. 271, 253 College Street, Toronto, Ontario, Canada M5T 1R5

First edition 1982

Second edition 1988

Library of Congress Cataloging-in-Publication Data

Dog & cat nutrition
Bibliography: P.
Includes index.
1. Dogs--Nutrition. 2. Cats--Nutrition. 3. Dogs-
Food. 4. Cats--Food. I. Edney, A. T. B.
II Title:
Dog and cat nutrition.
SF427.4.D63 1988 636.7'084 88-4048

British Library Cataloguing in Publication Data

Dog & cat nutrition.—2nd ed.
1. Pets: Dogs. Nutrition 2. Pets: Cats.
Nutrition
I. Edney, A.T.B.
636.7'0852

ISBN 0-08-035730-X Hardcover
ISBN 0-08-035729-6 Flexicover

Printed in Great Britain by A. Wheaton & Co. Ltd., Exeter

Foreword to Second Edition

The first edition of this book was published in 1982. Its objective was to make the subject of small animal nutrition accessible by making it readable. It set out to do this by taking readers from the basics of digestion and absorption, through nutrient needs and how they are met by the various foods available. The book also embraced more complex matters such as feeding animals which are ill, the special needs of growing dogs, orphaned puppies and individuals doing very hard work. A unique chapter on the evaluation and validation of prepared foods was included at the end. The second edition of *Dog & Cat Nutrition* has similar objectives but approaches them in a slightly different way.

In recent years the science and practice of small animal nutrition has moved on a considerable amount. With it, the general awareness of foods and feeding has grown enormously. This is especially true of the impact diet has on the management and prevention of a wide variety of canine and feline diseases.

This second edition, of what has become a standard work, takes the subject on to a greater stage of development in the light of recent advances. Much of this work has been carried out at the Waltham Centre for Pet Nutrition in the United Kingdom. As Europe's leading experts in this area, the staff of graduate scientists has been responsible for all of this edition. The format of the book has been simplified, although the breadth and depth of the coverage is increased. The section on small animal clinical nutrition has been greatly expanded in the light of current knowledge.

Anyone working their way through this book, whatever their state of knowledge on the subject at the start, can be sure of developing a reasonably sound grasp of a subject which has hitherto been given scant attention in any authoritative and readable form. The fact that this is so is due in no small measure to the gentle but persuasive and skilful help of our editor, Andrew Edney. All the Waltham authors wish to express their appreciation of his often unthanked industry.

Ian Burrows
Head, Waltham Centre for Pet Nutrition

Foreword to First Edition

Of the many books on nutrition, few relate to pet animals. This is surprising as, in the United Kingdom alone there are around 10 million dogs and cats. In the whole of the western world the dog and cat population is of the order of 150 million, virtually all of which need to be fed every day.

Nutrition is difficult to make readable, in spite of the obvious practical nature of the subject. Much of what is available is rather imprecise and in some cases has a liberal component of folklore. More accurate information tends to be less accessible to a general readership. There is it seems a need for a practically-orientated and above all readable account of the nutrition of dogs and cats. This book is meant to provide such a work for student veterinarians and animal nurses, those breeders of dogs and cats and all others who take a deeper interest in the feeding of their animals.

The book is structured so that the reader can progress from an understanding of the nutritional needs of dogs and cats, to the foods which are used to meet those needs. The middle section of the work is devoted to feeding in special circumstances, such as hard work, stress, orphaned puppies and various illnesses. The final chapter is included to explain how prepared foods are evaluated and validated, information which is difficult to find elsewhere. Only key references are included in the text, but a more comprehensive bibliography is given at the end of each chapter as a guide to further reading.

The contributors to this book are drawn from workers at the Animal Studies Centre in Leicestershire, and four of the world's leading authorities on dog and cat nutrition. These include Professor D. S. Kronfeld, who has kindly provided an account of his unique studies of sledge dogs as an investigation into the needs of animals doing very hard work.

The editor is pleased to express sincere thanks to Christina Loxley of the Animal Studies Centre and Dorothy Howard of Waltham for typing the manuscript and to Alison Wearne of Gwynne Hart & Associates, as well as John Lavender of Pergamon Press for progressing the work.

A. T. B. EDNEY

Contents

List of Contributors

Ian E. Burrows MSc PhD:
Ian Burrows graduated from the University of London with a degree in chemistry, followed by a PhD in biochemistry as a lecturer at Bradford University. In 1968, as Head of biochemistry at the Huntingdon Research Centre interests in human and animal nutrition were followed. Dr Burrows joined Pedigree Petfoods in 1972 and after an extensive career in the Research and Development Department is now Head of the Waltham Centre for Pet Nutrition.

Ivan H. Burger BSc PhD:
Ivan Burger graduated in physiology and biochemistry from Southampton University in 1968. He obtained his PhD at the Leatherhead Food Research Association in collaboration with Surrey University. In 1973 he joined the Waltham Centre for Pet Nutrition as Nutritional Biochemist and is currently Food Safety Co-ordinator and Pet Ownership Studies Manager.

Peter J. Markwell BSc BVetMed MRCVS:
Peter Markwell graduated from the Royal Veterinary College, University of London in 1981 also having obtained a degree in neuro-anatomy during the course. He entered small animal practice for 3 years. This was followed by a lectureship in the Department of Animal Husbandry at the Royal Veterinary College from 1984–85 before joining the Waltham Centre for Pet Nutrition as Veterinary Adviser. He is currently Senior Nutritionist at the Centre.

Anna L. Rainbird BSc PhD:
Anna Rainbird graduated from Queen Elizabeth College, University of London in 1980 with a degree in nutrition. In 1983 she obtained her PhD from the University of Reading after studying the effects of dietary fibre in pigs. Post-doctoral studies with both humans and pigs followed in both Reading and London. She joined Pedigree Petfoods in 1985 as Dog Nutritionist.

Andrew Edney BA BVetMed MRCVS:
Andrew Edney graduated from the Royal Veterinary College in 1958. He

spent the next 10 years in general veterinary practice and joined Pedigree Petfoods as Veterinary Adviser in 1968. He took an Arts degree in 1973 and was Secretary and President of the BSAVA and is currently Junior Vice-President of the World Small Animal Veterinary Association. After an early retirement he now acts as a freelance consultant, author and editor.

CHAPTER 1

Food Fit for Cats and Dogs

IAN E. BURROWS

INTRODUCTION

Animals feed in order to obtain all the nutrients and the energy necessary to sustain a healthy life and for successful procreation. Feeding any pet or companion animal should also be an occasion enjoyed by both the animal and the owner. This can only be achieved when suitable foods are offered which are readily eaten by the animal, and which provide it with a balanced diet. One of the major preoccupations for the owner of a pet or companion animal is keeping it fit and healthy. Dog and cat owners are no exception to this, and as with the owners of other companion animals, their prime concern is finding the right food to keep their particular pet fit.

Any consideration of what constituents make up a 'proper' diet for an animal must take into account both its lifestyle, and its lifestage. Most dog and cat owners keep their animals essentially as pets, but some expect them to perform other duties, for example, to hunt, race, keep down vermin, pull loads or for herding. All of these activities demand different types of diet, and therefore the food ingredients and the proportions from which they are composed will also differ. Dogs in particular have a large size range and both dogs and cats gestate and feed their young. Individuals may be growing old, pregnant or simply maintaining themselves as a normal healthy individual. Each lifestage makes it own particular demands on a diet, thus a puppy has a different nutrient need to an adult, or a pregnant queen to a lactating one. Each of these lifestages, or an animal with a particular lifestyle, needs a diet which is balanced in order to ensure that it remains fit and healthy. A balanced diet can be defined as that mixture of ingredients which provide all the energy and essential nutrients needed to maintain the animal in health appropriate to its lifestyle and lifestage. The question then, is to determine what foods can be used to provide dogs and cats with a balanced diet.

EVALUATION

To any owner of a cat or dog the most rewarding service that may be performed is to keep their companion healthy and fit, so that a long and happy life can be enjoyed. Probably the single most important aspect of achieving those ends lies with the diet that is provided. It is therefore essential that such diets have undergone an evaluation which guarantees performance. To fully evaluate the nutritional adequacy of any food or diet for an animal is no easy matter. This is particularly true for cats and dogs where the food is primarily intended to keep the animals fit and healthy throughout their lives. This demands a knowledge of the relationship between diet and health which is often far from being understood. Yet in many ways the nutritional feats expected of foods for cats and dogs often exceed those of both human food and farm animal feed. The reasons for this are not difficult to find, humans by and large have free access to a wide variety of foods, whereas cats and dogs are usually restricted to what their owners give them. Farm animals are fed for quite different reasons than pets: they are rarely expected to live out their natural lives. Their feed is designed to produce so much weight, or milk, or eggs, relative to time and costs. Health and efficiency obviously go together, but a lifetime of happy and efficient biological functioning is only expected for pets.

For any species the only effective method for evaluating and validating a diet is to feed it to the target animal, and to make a series of objective measurements relative to biological performance. Diets may be arrived at by trial and error (i.e. experience) but scientific guarantees of nutritional adequacy demand resources usually only available to professional nutritionists. Many leading manufacturers of prepared pet foods maintain facilities for carrying out complex and demanding nutritional assessments of their products. From such manufacturers the wide variety of products available may be used confidently and safely in the knowledge that their quality and nutritional status are assured.

VALIDATION

The basis of any validation of the nutritional adequacy of foods for cats and dogs must embrace a knowledge of the animal's requirements for specific nutrients and to match those against nutrient content of the food in question and its biological performance. Additionally it is important to ensure that the food is safe, i.e. contains no toxic elements, and that it is palatable. It is worth restating, uneaten food is nutritionally worthless, whatever its nutrient content.

TOXICITY

The supply routes of raw materials and the manner in which they can be contaminated is far too complex to be dealt with here, but the importance of keeping food free of toxins cannot be overstated.

The only certain way of ensuring that foods contain no toxic element is to possess a detailed knowledge of the history and source of all the individual ingredients. This however is usually impractical, the major guarantees are associated with purchasing raw materials from reputable sources, and the many and complex items of legislation which restrict, control and direct the quantities of additives and foreign bodies which may be present. Not all toxic substances are man made, many of the most virulent are natural in origin, and are often a feature of the raw material. Usually these features are removed by the farmer or processor. Toxins may also arise as a result of bad storage, and are due to outgrowths of bacteria or fungi. Such toxins are eliminated by good housekeeping practice, again it is essential to use reputable suppliers who understand the nature and quality of the raw materials which they handle. In addition to the protection of reputable supply and legislation, most leading manufacturers of food for dogs and cats operate elaborate quality control and screening procedures for both finished products and raw materials. Such procedures offer levels of guarantee for product safety, otherwise not easily ensured.

PALATABILITY

A book on nutrition cannot adequately deal with palatability, but again, the importance of highly palatable food cannot be over-emphasized. The palatability of food is a complex subject involving a knowledge of the factors affecting appetite and behaviour, as well as an understanding of the taste, smell and texture of food and their interrelationships. Making products which are consistently well eaten over extended periods requires a great deal of expertise and experience. Again reputable manufacturers of prepared foods for dogs and cats have developed objective measures for assessing palatability in order to ensure that any given recipe offers the consistent level of palatability expected by the animal, and the owner.

Individual dogs and cats, just like humans, have sharply individual preferences. To accommodate these tastes, large numbers of individual animals, often in the home environment, are studied for their likes and dislikes. In this way recipes can be developed which consistently give enjoyment both to the pet and the owner. These tests, which basically study intake under defined conditions, give rise to a large amount of complex statistical data, the interpretation of which give a measure of relative palatability, in terms of preference and acceptance.

Both dogs and cats enjoy good quality food, and find low quality foods less attractive. This may influence intake and result in 'problem' feeders, or even give rise to nutritional problems associated with poor intake. Most dogs and cats enjoy change and so benefit from variety in their choice of food. However, sudden changes of food type may produce digestive upsets. This is minimized by many manufacturers who produce variety within brands, such that the novelty of a new taste may be enjoyed without a disturbance of digestion.

BIOLOGICAL PERFORMANCE

The motivation for evaluating the biological suitability of cat or dog food is to provide the animal with a diet which is compatible with its health, and which satisfies its owner that the best is being provided. For the manufacturers of commercially prepared foods, it is self evident that it is in their best interests to ensure that their products satisfy their clients, and the purchaser. However, in addition to commercial interests, there are various areas of legislation (Appendix I) governing nutritional performance, and more importantly, the pet food industry promulgate and adhere to various voluntary codes of practice. In the space available it is not possible to give a comprehensive review on how each country considers the problem of biological evaluation. Therefore most emphasis has been placed on the North American and EEC situations.

To make a biological assessment of the nutritional quality of a food requires certain criteria to be carefully defined. For example a food may be positioned as being 'complete for all stages of life' whilst others may be 'complete for adult maintenance' or are for growth or pregnancy. The definition of terms like 'complete' or 'balanced' are often a source of confusion. However comprehensive definitions, used extensively by the pet food industry are provided by many national, official bodies. Amongst these are those used by the Association of American Feed Control Officials (AAFCO 1987) and the UK Feedingstuffs Regulations (1982) illustrated in Appendix II.

NUTRITIONAL TRIALS

Details of trials to determine the nutritional efficacy of a food or a diet for dogs and cats are given in Appendix III. All such trials ultimately depend on a knowledge of the specific nutrients which are required by dogs and cats. The major and most authoritative source of such information is published by the National Research Council (NRC), *Nutrient Requirements of Dogs* (1985) and the *Nutrient Requirements of Cats* (1986). These are occasional publications of an official body of the American National Academy of Science and represent, at the moment of publication, a summary of the expert knowledge then publically available from all sources around the world. The NRC guidelines for recommending the basic nutrient levels in food for both cats and dogs are based on the satisfactory breeding, rearing and maintenance of normal healthy animals. These guidelines therefore provide target values for which nutritional claims are made. Clearly the information and tables provided can only form a set of guidelines since inevitably between one publication and the next, some of the information becomes out of date due to the refinements of new knowledge.

DIGESTIBILITY

The simplest possible nutritional test of a food is to analyse it chemically for specific nutrients such as protein, carbohydrate, fat, trace minerals, vitamins, amino-acids etc. This then gives an idea of how much of any given nutrient is present in the food. What such a procedure cannot do is to give any estimate of the biological value of the food. For the nutrients to be of any use to the animal they must be biologically available, i.e. the animal's digestive system must be able to extract the nutrient from the food and incorporate it into its metabolic pool. To obtain an idea of how biologically available the nutrients are, i.e. an estimate of digestibility, the analysed food is fed to an adult dog or cat in measured quantities over a given fixed period, usually 2 weeks. During that period all voided urine and faeces are collected and ultimately analysed for the same nutrients as the food. The difference between the quantity of nutrients taken into the animal in the food, and that voided, gives a measure of the amount of each nutrient retained or utilized by the animal. This then serves as an index of digestibility. This index can then be used to estimate if the food, and what quantity of food will maintain a normal healthy animal, by supplying all the essential nutrients and energy required by that particular individual.

GESTATION AND LACTATION

The most exacting test for any food is to feed it through the gestation, lactation and rearing period of young animals. For a dog or cat, two groups of female animals are kept, one group fed on a test diet, and the other group on a control, or standard diet. Both groups are fed through the mating, pregnancy and lactation period until the puppies or kittens are 6 to 7 weeks old. Bodyweights of dams and individual young are recorded, as are food intake and litter size. Veterinary health checks are made routinely and at intervals during the trial assessments to ensure that all the animals are growing normally and maintaining their health.

GROWTH

During the growth period, young animals are making particular demands on food; often the specific nutrient requirements of growing animals being different to those needed for adult maintenance. Young kittens or puppies matched for weight, size, sex and genetic background are fed in two groups, one group fed the test diet, the other the reference or standard diet. The daily food consumption is measured, and the weekly weight gain for each individual is recorded throughout the trial. The growth period is followed from 4 to 6 months following weaning, during which extensive health checks are made. Statistical comparisons can then be made at the end of the trial in order to assess the potential of the test diet to sustain normal healthy growth.

MAINTENANCE

The acid test for any food is how well it can sustain the needs of animals, and to keep them healthy over very long periods. To carry out such tests it is necessary to follow the feeding of that food over several generations. Clearly to achieve this requires the ability to study large numbers of animals over very extended periods of time. Information gleaned from such trials gives invaluable insights into the most difficult of nutritional equations, namely the relationship between diet and health.

Cats and dogs have been associated with humans for several thousand years, and although this is only a fleeting instant in terms of geological or evolutionary time, these very different species have adapted well to living together. Superficially, because of this symbiosis it may appear that dogs, cats and humans have similar dietary needs. Certainly there is a very considerable overlap in what foods they will each eat, and as a general rule what is suitable for humans, will to some degree be suitable for a dog or a cat. However what they need to eat, or what constitutes a balanced diet, is very different for each of them.

These differences are undoubtedly due to the disparate evolutionary niches attained by dogs and cats. Dogs have a common ancestry with wolves, and still retain strong social or pack instincts, whilst cats are essentially solitary animals. Dogs and cats both belong to the mammalian order Carnivora, but there the similarity ends. Dogs, although carnivorous in habit, are able to obtain nutrients from a wider range of foods, e.g. vegetable material such as roots, nuts and fruits, as well as animal tissues. Dogs do not however readily digest plant material, especially celluloses and hemicelluloses, but although it is conceivable that a dog could be maintained on a carefully selected vegetarian diet, it is almost certain that a cat could not. Cats are obligate carnivores and can only obtain all the essential nutrients that they need from animal tissue; they therefore have a much more restricted diet than dogs.

As a consequence of this difference in diet, not only do dogs and cats have differing needs for various nutrients such as vitamins, amino-acids, minerals and protein, but they also have different physiological mechanisms for extracting those nutrients from foods. Dogs often have enzyme systems which allow them to synthesize nutrients in their own body from nutrient precursors obtained from food. In many instances cats lack these enzymes and so must obtain the completed nutrient from their food. Good examples of these differences are taurine and arachidonic acid, which cats have only a limited ability to synthesize, and which are simply not present in plant material. It is easy to see that the foods which make up a cat's diet must be different from those for a dog.

Nutrition is not the only characteristic of a food which is important to either an animal or its owner. Feeding a pet has a high social content, therefore it is important that the animal eats its food eagerly, and with enjoyment. Food must have the right smell, flavour, and texture, and just as important to the owner, the food should be acceptable in a domestic environment. Its preparation and storage should be convenient, it should not have an objectionable smell and it should be easy to handle and serve. As important, it should be safe, i.e. free from toxic contaminants and it should not be spoiled, or made bad, by bacterial or fungal outgrowths. Most of these latter problems are dealt with by cooking or otherwise processing the food.

The palatability and acceptance of food to dogs and cats are attributes easy to recognize, but hard to describe. No matter how good the nutrition of a food is, if it is uneaten then it fails to deliver its nutrition, so good palatability is crucial. The total diet of a dog or cat may consist of a single type of food, fed on a daily basis, and provided it represents a balanced diet it can adequately maintain the animal in health. Like humans, dogs and cats enjoy variety, which can easily be achieved by feeding a mixture of foods, which is also the easiest way of ensuring a good mix of nutrients. Single foods rarely offer nutrients in a correct balance.

BIBLIOGRAPHY

AAFCO (1986) Association of American Feed Control Officials, Official Publication. Copies may be obtained from Donald H. James, Department of Agriculture, State Capital Building, Charlestown, W. Virginia, U.S.A.

Bradshaw, J. W. S. (1986) Mere exposure reduces cats' neophobia to unfamiliar food. *Anim. Behav.* **34**, 613–614.

Houpt, K. A. and Smith, S. L. (1981) Taste preferences and their relation to obesity in dogs and cats. *Can. Vet. J.* **22**, 77–81.

Houpt, K. A. and Wolski, T. R. (1982) *Domestic Animal Behaviour for Veterinarians and Animal Scientists.* Iowa State University Press, Ames, Iowa, U.S.A.

Kendall, P. T. (1981) Comparative evaluation of apparent digestibility in dogs and cats. *Pro. Nutr. Soc.* **40**, 245A.

Kitchell, R. L. and Baker, G. C. (1972) Taste preference studies in domestic animals. *University of Nottingham Conference for Feeding Manufacturers,* No.6, pp. 157–202. Churchill Livingstone, London.

Mugford, R. A. (1977) External influences on the feeding of carnivores. *The Chemical Senses in Nutrition.* Academic Press, London.

NRC (1985) *Nutrient Requirements of Dogs,* Report National Research Council, National Academy of Sciences, Washington, U.S.A.

NRC (1986) *Nutrient Requirements of Cats,* Report Nutritional Research Council, National Academy of Sciences, Washington, U.S.A.

Thorne, C. J. (1982) Feeding behaviour in the cat — recent advances. In Recent Advances in Feline Nutrition, Waltham Symposium No.4. *J. Small Anim. Pract.* **23**, 555–562.

Thorne, C. J. (1985) Cat feeding behaviour. *Pedigree Digest* **12**, 4–6.

CHAPTER 2

A Basic Guide to Nutrient Requirements

IVAN H. BURGER

Like all other living creatures, dogs and cats require food to stay alive and healthy. Food may be defined as 'any substance which is capable of nourishing the living being'. A more complete description is that food is any solid or liquid which when swallowed can supply any or all of the following:

(a) energy-giving materials from which the body can produce movement, heat or other forms of energy;
(b) materials for growth, repair or reproduction;
(c) substances necessary to initiate or regulate the processes involved in the first two categories.

The components of food which have these functions are called nutrients, and the foods or food mixtures which are actually eaten are referred to as the diet. The main types of nutrients present in foods are:

Carbohydrates — these provide the body with energy and may also be converted into body fat. This group includes simple sugars (such as glucose) and larger molecules (such as starch) which consist of chains of the simpler sugars.

Fats — these provide energy in the most concentrated form, releasing about double the amount of energy per unit weight than either carbohydrates or protein. Fats aid in the absorption of the fat-soluble vitamins and supply types of fat usually referred to as the essential fatty acids (EFA). These, as their name suggests, are required for certain important body functions and are as important as individual vitamins or minerals. The EFA will be discussed in more detail later in this chapter.

Proteins — these are important because they provide amino-acids which are involved in the growth and repair of body tissue. The component amino-acids can also be metabolized to provide energy.

Minerals and trace elements — the 'major minerals' are substances like calcium and phosphorus which are used in growth and repair, and make up most of the skeletal and tooth structure. This category also includes substances required in smaller quantities such as iron, copper and zinc. The latter group are usually referred to as trace elements.

Vitamins — these help to regulate body processes and are usually considered as two categories, the fat soluble and water soluble groups. In the former are vitamins A, D, E and K; the latter group includes vitamins of the B complex (such as thiamin) and vitamin C.

The other important constituent of food is water and although this is not generally regarded as a nutrient, it is essential to life. Water balance is discussed in greater detail in the next chapter. The need for water is second only to the need for oxygen, the other vital element not included in the list above.

Hardly any foods contain only one nutrient: most are complex mixtures which consist of a variety of carbohydrates, fats and proteins together with water. Minerals and vitamins (especially the latter) are usually present in much smaller amounts.

REQUIREMENTS AND RECOMMENDATIONS

An adequate intake of nutrients is essential for the health and activity of the animal, but how much is adequate? Compared with the requirements of the adult dog or cat, there are additional needs for the more demanding stages of the lifecycle such as growth, pregnancy and lactation. In the case of the dog and cat, it is possible to investigate their needs for nutrients and to obtain more precise values than is possible for man. The minimum quantity of an individual nutrient which must be supplied each day for proper body metabolism is usually referred to as the minimum daily requirement (MDR). The National Research Council (NRC) of the National Academy of Sciences of the USA has compiled a list of the minimum nutrient requirement of dogs and cats for growth (Appendices IV and V) and these will be referred to frequently throughout the book. It must be remembered, however, that the NRC values are *minima* and appropriate allowances must be made for individual variation, physical activity, breed, weight, sex and stage of development. Furthermore, there are other factors which must be taken into account, in particular effects on the availability of nutrients in foods and these will be discussed later in this chapter. In view of all these considerations, it is more practical to use the NRC data to derive values for a recommended daily allowance (RDA) as a preferred guide for nutritional adequacy. For any animal the RDA is designed to ensure that the needs of virtually all the normal healthy individuals in the population are covered. It follows that the RDA will always be in excess of MDR (except for energy which is discussed below) and actual experimentally determined requirements will be less than recommended intakes. It also follows that a diet may contribute *less* than the RDA, but still provide an adequate nutrient intake for certain animals.

An equally important aspect is the application of RDA (or MDR) to a food or mixture of foods i.e. the diet. Requirements will initially be assessed as a quantity of nutrient ingested by the animal and will usually have units of intake per kg bodyweight per day. But ultimately the most useful and relevant way to express this value is as a concentration in the diet. This raises the question of the quantity of different types of foods eaten by different animals. Foods have different compositions (from canned to dry) and animals, particularly dogs, show a wide variation in size from breed to breed. The link between these variables is the energy content of the diet.

ENERGY

Energy is different from other nutrients, in that appetite normally controls intake and keeps it close to requirements. Intakes in excess of requirements are undesirable and eventually lead to obesity. The energy content of a diet is derived from carbohydrates, fats and protein, and the amount of each of these nutrients in a food will determine its energy content. Water has no energy value, so the energy density of food varies in an inverse relation to its moisture content. Energy is usually expressed in terms of kilocalories (kcal) where 1 kcal is defined as the quantity of heat required to raise the temperature of 1 kg of water by 1 centigrade degree. A more recent convention is to express energy in terms of the kilojoule (kJ) which is more difficult to define in familiar terms and is based on a mechanical or electrical equivalent of heat. For the purposes of this discussion it is necessary only to realize that 1 kcal is equivalent to about 4.2 kJ. The body obtains energy by oxidizing ('burning') food but, unlike the burning process in a boiler or engine, the energy is released gradually by a series of complex chemical reactions, each regulated by an enzyme.

Enzymes are special proteins which control the rate of chemical reactions and more importantly, enable these complex changes to take place in the relatively mild conditions of the body. To bring about the same changes in a typical industrial process would require much more extreme conditions of temperature and pH or highly reactive ingredients. Many enzymes require the presence of vitamins or minerals to function properly and this aspect will be discussed in more detail when these nutrients are considered.

Dogs and cats, like all animals, are unable to extract all the energy from food. Energy intake is therefore considered at three different levels: gross energy (GE), digestible energy (DE) and metabolizable energy (ME). Gross energy is the total energy released by complete oxidation of the food and is usually measured by burning it in an atmosphere of pure oxygen in an instrument (calorimeter) which accurately measures the heat released on combustion. Although a substance may have a high GE content, it is of no use to the dog and cat unless the animal is able to digest and absorb it. The amount which is digested and absorbed is known as DE and equals GE minus faecal losses. Some of the absorbed food may only be partially available to the tissues, the remainder being lost via the kidneys in the urine. The energy which is ultimately utilized by the tissues is known as ME and is calculated as DE

TABLE 1

Calculation of the metabolizable energy content of pet foods from chemical analysis

Dog foods	$ME = 3.5 \times P + 8.5 \times F + 3.5 \times CHO$
Cat Foods	
canned	$ME = (3.9 \times P + 7.7 \times F + 3.0 \times CHO) - 5$
semi-moist	$ME = 3.7 \times P + 8.8 \times F + 3.3 \times CHO$
dry	$ME = 0.99 \times (5.65 \times P + 9.4 \times F + 4.15 \times CHO) - 126$

P = protein content; F = fat content; CHO = carbohydrate content, all expressed as g/100g. ME = metabolizable energy content of the food, expressed as kcal/100g. (from NRC, 1985 and NRC, 1986)

minus urinary losses. The DE and ME contents of foods depend both upon their composition and upon the species which eats them. For example, the digestive system of the dog seems more efficient than that of the cat (this may be partly because the dog's digestive system is proportionately longer than the cat's). Therefore the same food fed to dogs or cats could yield different digestibility values. There will also be variations between individual animals in their own metabolic efficiency. So the only way to obtain a meaningful measurement of the ME content of a food is to feed it to as large a group of dogs or cats as possible and measure energy (using the calorimeter technique) in food, faeces and urine. This technique, although perfectly feasible, is time-consuming and costly and is not possible without access to specialized animal facilities.

Therefore over the years a simple formula has been developed which gives a reasonable approximation of the ME in a food from its carbohydrate, fat and protein contents, allowing for the losses in absorption and efficiency. The factors originally used were those derived from studies in man. More recently the NRC has published more accurate figures compiled from feeding studies in dogs and cats. These are summarized in Table 1. Studies conducted at the Waltham Centre for Pet Nutrition (WCPN) in which DE measured *in vivo* was compared with ME predicted from factors suggest that the new NRC values give a good estimate of the energy available to the dog and cat from typical commerical pet foods.

Energy is used to perform muscular work, processes such as breathing and physical activity to maintain body temperature. Like man, the dog and cat maintain their body temperature at around 40°C, normally well above the environmental conditions, and large amounts of energy are required to achieve this. Thus the first requirement of the animal from its diet is energy. The energy density of the diet must be high enough to enable the dog or cat to obtain sufficient calories to maintain energy balance. This is the principal factor determining the quantity of food eaten each day and thus the amount of each nutrient ingested by the animal. Therefore in the following sections nutrient requirements are usually expressed in terms of the ME concentration so that the values are applicable to any type of food or diet regardless of its water content, nutrient content or overall energy value.

NUTRIENT FUNCTIONS AND REQUIREMENTS

Carbohydrates

There is no known minimum dietary carbohydrate requirement for either the dog or cat. Based on investigations in the dog and with other species it is likely that dogs and cats can be maintained without carbohydrates if the diet supplies enough fat or protein from which the metabolic requirement for glucose is derived. For example, it has been reported (Romsos *et al.* 1981) that the consumption of a high-fat carbohydrate-free diet by bitches during gestation substantially reduced the survival of their puppies compared with a control group receiving a diet containing 44% ME as carbohydrate. The effect was attributed to a severe hypoglycaemia in the former bitches at whelping. Nevertheless a study has been conducted at WCPN in which two diets were compared with regard to their ability to support Beagles and Labradors through pregnancy and lactation (Blaza *et al.* 1988). One diet supplied no available carbohydrate whereas the other contained 11% of energy as carbohydrate. No differences in performance were seen between these two treatments — both supported normal pregnancy and lactation. The contrast between these two studies can probably be explained by differences in the protein levels in the two diets. In the latter investigation the protein level was much higher — high enough to supply an adequate level of glucose. Thus while carbohydrate is *physiologically* essential it is not an indispensable component of the diet.

The carbohydrate source used in both of these studies was cooked starch and there is little doubt that this substance is readily digested by both dogs and cats. Individual disaccharides (i.e. containing two sugar units) such as sucrose (cane sugar) and lactose (milk sugar) are less well tolerated. The ability to metabolize these sugars is governed, respectively, by the amounts of the enzymes beta-fructofuronidase (sucrase) and beta-galactosidase (lactase), present in the intestine. Sucrase and lactase activities are certainly present in adult dogs and cats, although they are known to be higher in kittens and to decline with increasing age. If adult or young dogs and cats are suddenly given *large* amounts of sucrose or lactose (for example a large bowl of milk) they may exhibit diarrhoea which is due partly to osmotic purgation and partly to bacterial fermentation (in the large intestine) of carbohydrate escaping digestion. Despite this, small quantities of these carbohydrates (say 5% of total calories) can be well tolerated by most animals although there will obviously be variations in the efficiency of individual animals in utilizing these substances.

Some work with dogs (Kronfeld *et al.* 1977) has suggested that a carbohydrate-free, high-fat diet actually confers some advantages for prolonged strenuous running in racing sledge dogs, compared with diets containing up to 38% of calories as carbohydrate. These advantages included a higher oxygen-carrying capacity in the form of more red blood cells and haemoglobin. However, for normally active dogs and cats the inclusion of 40–50% of calories as dietary carbohydrate is unlikely to represent any disadvantage compared with a total fat and protein diet.

Fat

Dietary fat serves as the most concentrated source of energy in the diet and lends palatability and an acceptable texture to dog and cat foods. Like carbohydrates, fats are compounds of carbon, hydrogen and oxygen. Chemically, food fats consist mainly of mixtures of triglycerides where each triglyceride is a combination of three fatty acids, joined by a unit of glycerol. The differences between one fat and another are largely the result of the different fatty acids in each. There are many different fatty acids found in foods and their chemical structures are characterized by the number of carbon atoms and double bonds. Saturated fatty acids have no double bonds, whereas the unsaturated variety have one or more; those containing more than one double bond are referred to as polyunsaturated. Most fats contain all of these types but in widely varying proportions.

It is difficult to give a precise requirement for total dietary fat for dogs and cats. The only demonstrable need for fat is as a provider of EFA and a carrier of the fat soluble vitamins. These functions will determine the requirement for fat together with the need to provide a certain level in the diet to achieve the necessary energy density and palatability. There are three recognized EFA, linoleic, α-linolenic and arachidonic acids, all of which are polyunsaturated. Because of the complex nature of these compounds it is usual to designate their structure by the number of carbon atoms and double bonds they contain; thus linoleic acid which contains 18 carbon atoms and two double bonds is written 18:2. The EFA cannot be synthesized by the body and are therefore essential nutrients which must be supplied in the diet. Linoleic and α-linolenic acids are the parent compounds from which the more complex, longer chain compounds (derived EFA) can be synthesized by the body. EFA are important for the general health of the animal and are involved in many aspects of health including skin and coat condition, kidney function and reproduction. It is in the formation of the EFA that an important difference between the dog and cat emerges, a contrast that is repeated for other nutrients and in which the cat is atypical, in that the dog follows the pattern for most other mammals but the cat does not. It has been reported that cats have only a limited ability to convert the parent EFA into the longer chain derivatives (Rivers, 1982); the lion appears to be similar in this respect. As a result cats require a pre-formed dietary source of 20:3 or 20:4 (arachidonic) acids which in practical terms means a requirement for EFA of *animal* origin.

In an elegant study of EFA in cats MacDonald *et al.* (1984a) concluded that dietary linoleic acid at 2.5% of energy was probably adequate and that, given an ideal level of linoleate, the arachidonic acid requirement was not less than

0.04% of energy. However, the interrelationships between these two compounds means that a higher level of arachidonic acid in the diet will spare the need for linoleic acid. Conversely, the minimum requirement for arachidonic acid would be much more than 0.04% if linoleic acid was below the ideal level or absent from the diet altogether. In practical terms the EFA requirements of the cat are met by a combination of linoleic and arachidonic acids (the former being more widely available than the latter) from a blend of vegetable and animal oils and fats in the food.

Protein and amino-acids

All proteins are compounds of carbon, hydrogen and oxygen but unlike carbohydrates and fats they always contain nitrogen. Most proteins also contain sulphur. Proteins are very large molecules which consist of chains of hundreds (or perhaps thousands) of much smaller sub-units called amino-acids. Although there are only about 20 amino-acids used in the composition of proteins, the variety of sequences in which they can be arranged is almost infinite and this results in the wide variety of proteins which occur in nature. Cats and dogs need dietary protein to provide the specific amino-acids that their tissues cannot synthesize at a rate sufficient for optimum performance. These amino-acids are then reformed into new proteins which are essential constituents of all living cells where they regulate metabolic processes (in the form of enzymes), provide structure and are therefore required for tissue growth and repair. Amino-acids can be conveniently divided into two classes: essential (indispensable) and non-essential (dispensable). As their name suggests, the essential amino-acids cannot be made by the body in sufficient amounts and must therefore be present in the food. The non-essential amino-acids can be made from excesses of certain other dietary amino-acids, although, as components of body proteins, they are as important as the essential varieties. Essential amino-acid requirements for kitten and puppy growth and for adult dog maintenance have been the subject of much intensive investigation over the last decade. The NRC has summarized the available data on amino-acid requirements and these are shown in Appendices IV and V. Requirements for the adult cat have yet to be determined and we are currently investigating this area at the Waltham Centre for Pet Nutrition (WCPN). Earlier work at WCPN has shown that when all essential amino-acids are present at more than adequate concentrations, about 10% protein calories are required to maintain adult cats in protein (nitrogen) balance (Burger *et al.* 1984). This value is higher than corresponding figures for the dog and is another example of nutritional differences between dogs and cats. The higher protein requirement of the cat does not appear to be due to an increased requirement for essential amino-acids but rather a need for more total protein i.e. indispensable amino-acids or protein nitrogen. This, in turn, appears to be due to the cat's inability to adjust amino-acid breakdown even when receiving a low protein diet (Rogers and Morris, 1982). The cat is also unusual in its dependence on the amino-acid arginine. Arginine deficiency in the cat rapidly results in severe adverse effects because of an inability to metabolize nitrogen compounds (via the urea cycle), which then accumulate in the blood-stream as ammonia (hyperammonaemia) and in serious cases can lead to death within several hours. It seems that there is no other essential dietary component (including water) whose deficiency has such a drastic effect upon the animal. The rapidity of the effects is second only to a lack of oxygen. This unique requirement appears to be due to an inability to synthesize the amino-acid ornithine (also a component of the urea cycle) since the latter protects cats against the adverse effects of arginine deficiency (Morris and Rogers, 1978). Although other animals require arginine for growth, in general they do not need it for adult maintenance. Those that do (like the dog) seem to be much less sensitive to a deficiency and have a much lower dietary requirement than cats.

Most if not all of the investigations into protein and amino-acid requirements of dogs and cats have been conducted using semi-purified or 'synthetic' diets, in which the protein level or amino-acid profile has been adjusted for the purposes of the study. In extrapolating these results to practical feeding or design of diets it is important to allow for several factors. The essential amino-acid profile of a given protein is of paramount importance. Few, if any, naturally occurring proteins would have the amino-acid content of a specially made test diet. Furthermore, availability or digestibility of proteins will vary from one source to another and from one animal to another. Animal proteins generally have a more balanced amino-acid profile and better digestibility than plant proteins. This whole subject represents a good example of the difference between a precise requirement, determined under carefully defined and controlled test conditions, and a recommendation which must apply to a very large number of animals eating a wide range of foods.

Despite these limitations, studies to determine the precise protein and amino-acid requirements of the dog and cat are important steps in refining the formulation of diets for these animals. Protein is a precious raw material and one which should be used as efficiently as possible.

Taurine

No discussion of the amino-acid requirements of the dog and cat would be complete without at least a brief explanation of the importance of taurine. Strictly speaking, taurine is an amino-sulphonic acid which is not part of the polypeptide chain of protein. It is an end-product of sulphur amino-acid metabolism and is normally produced from the sulphur-containing amino-acids, methionine and cystine. The particular importance of taurine in cat nutrition was first discovered only 13 years ago when Hayes *et al.* (1975) showed that taurine was an essential nutrient for the cat and a deficiency was associated with central retinal degeneration. Unlike dogs, cats cannot synthesize sufficient taurine to meet their needs, and the special sensitivity of the cat is heightened by its total dependence on taurine for the formation of bile salts. Unlike other species, it does not also use glycine for this purpose (see Chapter 3). The cat is therefore dependent on a dietary supply of taurine and the 'animal dependence' theme is again shown in this instance as taurine is found almost exclusively in animal-derived materials, little is present in plants.

Although the discovery of taurine function in the cat was centred on retinal function, more recent research suggests that the importance of taurine in cat nutrition extends beyond this area. Sturman *et al.* (1986) reported that a taurine-free diet fed to queens during gestation and lactation resulted in poor reproductive performance typified by frequent foetal resorption, low birth weight of kittens, poor survival and a reduced growth rate. Abnormalities in neurological function and skeletal growth also occurred. The latest information (Pion *et al.* 1987) suggests that taurine deficiency in cats is also linked with dilated cardiomyopathy—a degenerative disease of the heart muscle.

Minerals

Calcium and Phosphorus

Calcium and phosphorus are closely interrelated nutritionally and will therefore be discussed together. They are the major minerals involved in the structural rigidity of bones and teeth. Calcium is also involved in blood clotting and in the transmission of nerve impulses. The level of calcium in the blood plasma is crucial to these functions and is very carefully regulated. Phosphorus also has many other functions (more than any other mineral element), and a complete discussion of phosphorus metabolism would require coverage of nearly all the metabolic processes in the body. Phosphorus is involved in many enzyme systems and is also a component of the so-called 'high-energy' organic phosphate compounds. These are mainly responsible for the storage and transfer of energy in the body.

The ratio of calcium to phosphorus in the diet is perhaps of greatest importance. The optimum calcium to phosphorus ratios for dogs and cats are generally considered to lie between 1.2 and 1.4:1 and 0.9 to 1.1:1, respectively. Imbalance in this ratio, where calcium is much less than phosphorus, leads to a marked deficiency of calcium in relation to bone formation. There is also evidence that a very high ratio is harmful. The metabolism of calcium and phosphorus is closely linked with vitamin D and this will be discussed later in the chapter.

Potassium

Potassium is found in high concentrations *within* cells and is required for nerve transmission, fluid balance and muscle metabolism. A deficiency causes muscular weakness, poor growth and lesions of the heart and kidney. However, potassium is widely distributed in foods and naturally occurring deficiencies are extremely rare.

Sodium and Chloride

In contrast to potassium, sodium occurs mainly in the extracellular fluids, but like potassium, it is important for normal physiological function. With chloride, these substances represent the major electrolytes of the body water. Common salt (sodium chloride) is the most usual form of these two minerals added to food, so the dietary recommendation is often expressed in terms of sodium chloride. As with potassium, it is most unlikely that normal diets will be deficient in these two minerals. There is evidence that, in dogs, excessive sodium intakes can cause adverse effects linked to hypertension. In cats there is much less information available but it appears cats may not be as susceptible. In any event it seems advisable to set a maximum sodium level (especially for dog foods) to reduce the likelihood of any health problems.

Magnesium

Magnesium is found in the soft tissues of the body as well as in bone. Heart and skeletal muscle and nervous tissue depend on a proper balance between calcium and magnesium for normal function. Magnesium is also important in sodium and potassium metabolism and plays a key role in many essential enzyme reactions, particularly those concerned with energy metabolism. A deficiency of magnesium is characterized by muscular weakness and in severe cases convulsions. Nevertheless a dietary deficiency of magnesium is very unlikely. In contrast, very high intakes of magnesium in cats are associated with an increased incidence of the Feline Urological Syndrome (FUS). This is discussed in detail in Chapter 6.

Trace Elements

Iron

Iron is probably the best known trace element and much research has been carried out on its functions and requirements, particularly in the dog. Iron is a component of haemoglobin and myoglobin which play an essential role in oxygen transport; it is also an essential part of many enzymes (haem enzymes) which are involved in respiration at the cellular level, i.e. the oxidation of nutrients to form chemical energy. The aborption of iron is known to be influenced by a number of factors. Ferrous iron is better absorbed than ferric iron, and iron contained in foods of animal origin tends to be better absorbed than that from vegetable sources. Some recent evidence from studies in man suggests that the inclusion of soy protein in a diet reduces the absorption of iron and other trace elements (zinc and manganese) and it may be important to ensure that the concentration of iron in products containing high levels of soy protein is always above the recommended allowance.

A deficiency of iron results in anaemia with the typical clinical picture of weakness and fatigue. Conversely iron, like most trace elements, is toxic if ingested in excessive amounts. Iron toxicity in dogs has been extensively studied, and is associated with anorexia and weight loss. Of the iron salts investigated, ferrous sulphate was the most toxic, presumably because its absorption is high; iron oxide was far less dangerous, because its bioavailability is very low.

Copper

Copper is involved in a broad range of biological functions and is a constituent of many enzyme systems, including one which is necessary for the formation of the pigment melanin. Copper is very closely linked with iron metabolism and its deficiency impairs the absorption and transport of iron and decreases haemoglobin synthesis. Thus a lack of copper in the diet can cause anaemia even when the intake of iron is normal. Bone disorders can also occur as a result of copper deficiency and in this case the cause is thought to be a reduction in the activity of a copper-containing enzyme leading to diminished stability and strength of bone collagen.

Ironically *excess* dietary copper may also cause anaemia which is thought to result from competition between copper and iron for absorption sites in the intestine. Bedlington Terriers are known to display an unusual defect which results in toxic excesses of copper in the liver. The disorder results in hepatitis and cirrhosis and appears to be inherited. It has also been identified in other breeds including West Highland White Terriers and Dobermann Pinschers (Thornburg *et al.* 1985a,b). For these particular breeds of dog it is probably a good idea to exclude foods with high copper contents and to avoid the use of copper-containing mineral supplements.

Manganese

Although little is known about the specific manganese requirements of dogs or cats, a considerable amount of evidence has accumulated that this trace element is essential in animal nutrition and there is no reason to suppose that the dog and cat are any different in this respect. Manganese is known to activate many metal–enzyme systems in the body and is therefore involved in a wide variety of reactions. A deficiency of manganese is characterized by defective growth and reproduction, and disturbances in lipid metabolism. These effects, like those of copper deficiency, are probably caused by inactivation or malfunction of one or more of the enzyme-catalyzed reactions associated with these physiological processes.

Although manganese is reported to be one of the least toxic of the trace elements, toxicity has been reported in several species, including cats, where it caused poor fertility and partial albinism in some Siamese. One of the other effects of excess manganese is on haemoglobin formation where its action is thought to be similar to that previously described for copper, i.e. competition with iron at the absorption sites in the alimentary tract.

Zinc

The functions of zinc can be divided into two broad categories: enzyme function and protein synthesis. Zinc is required by both dogs and cats, but the zinc requirement is particularly affected by other components of the diet. For example, a high dietary calcium content or a vegetable protein-based diet can dramatically increase the zinc requirement and this latter effect may be related to that reported for iron absorption. Zinc availability is also decreased by the presence of phytic acid in the food. This compound is a complex organic molecule containing phosphorus which can bind trace elements such as zinc and thereby reduces their availability to the animal. Phytic acid and its derivatives (the phytates) are found particularly in cereals or related materials. In foods containing these, care must always be taken to ensure that the zinc concentration is adequate. Van den Broek and Thoday (1986) reported signs of zinc deficiency in dogs receiving cereal-based dry diets which contained zinc levels that were actually above the NRC minimum requirement.

Zinc deficiency is characterized by poor growth, anorexia, testicular atrophy, emaciation and skin lesions. Although all nutrients are important, the link between zinc and skin and coat condition makes this trace element particularly crucial for the pet animal. This is because a marginal deficiency may occur where the dog or cat is not obviously unwell but its skin or coat condition is sub-optimal and significantly detracts from the appearance of the animal. Zinc is relatively non-toxic. It interferes with the absorption and utilization of iron and copper (especially the latter) so the severity of the effects of high intakes of zinc is dependent on the levels of these other trace elements in the diet. With normal dietary contents of iron and copper, it appears that zinc concentrations up to eight times the minimum requirement will not produce adverse effects.

Iodine

The only recognized function of iodine is in the synthesis of the thyroid hormones which are released by the thyroid gland and regulate the metabolic rate of the animal. One of the factors which influences the output of the thyroid hormones is the availability of sufficient iodine. In the absence of the requisite amount the thyroid gland increases its activity in an attempt to compensate for the iodine deficiency. As a result the gland (which is located at the neck region of the animal) enlarges and becomes turgid, a condition known as goitre which is the principal sign of iodine deficiency. Nevertheless there are other factors which are important in the occurrence of goitre. These include infectious agents, naturally occurring substances in the diet (goitrogens) which inhibit the synthesis, release or general effectiveness of the thyroid hormones, and genetically determined defects in the enzyme systems responsible for the biosynthesis of these hormones.

In man severe reduction in thyroid activity (hypothyroidism) is often referred to as cretinism when it occurs in children, and myxoedema in adults. Hypothyroidism has been reported in dogs and iodine deficiency has also been observed in zoo felids and domestic cats. Clinical signs include skin and hair abnormalities, dullness, apathy and drowsiness. There can also be abnormal calcium metabolism and reproductive failure with foetal resorption. Excessive iodine intakes can be toxic. Hypothyroid cats given high doses of iodine (about 150 times the minimum requirement) were reported to show adverse effects which included anorexia, fever and weight loss (NRC, 1986). In other animals very large doses of iodine have been reported to produce acute effects similar to those of a *deficiency*. The high doses in some way impair thyroid hormone synthesis and can produce so-called iodine myxoedema or goitre.

Selenium

Ironically, attention was first focussed on selenium because of its toxicity. The discovery that it is an essential nutrient for mammals took place fairly recently, about 30 years ago. Any discussion of the biochemical role of selenium has to take into account the close interrelationship of this element with vitamin E and the sulphur-containing amino-acids methionine and cystine. The link with vitamin E is particularly important since one nutrient can 'spare' a deficiency of the other. Nevertheless it has been demonstrated in many animals, including the dog, that selenium cannot be replaced completely by vitamin E and has a discrete, unique function. Selenium is known to be an obligatory component of an enzyme called glutathione peroxidase which protects cell membranes against damage by oxidizing substances (notably lipid peroxides) which can be released by various metabolic processes in the body. Sulphur amino-acids are required to form the enzyme; vitamin E is thought to act within the membranes, preventing oxidation of the lipids. In this way the functions of these three nutrients are closely linked.

The interactions of selenium are obviously highly complex and much is still unknown about this substance. It may, for example, be involved in processes unrelated to its role as a component of glutathione peroxidase. It has been reported to protect against lead, cadmium and mercury poisoning and has even been implicated as an anti-cancer agent in both experimental and epidemiological studies. Selenium deficiency has many effects but one described in dogs is degeneration of skeletal and cardiac muscle. Effects of deficiency in other species include reproductive disorders and oedema.

TABLE 2
Summary of functions of some trace elements

Element	Involvement
Chromium	Carbohydrate metabolism, closely linked with insulin function
Fluoride	Teeth and bone development, possibly some involvement in reproduction
Nickel	Membrane function, possibly involved in metabolism of the nucleic acid RNA
Molybdenum	Constituent of several enzymes, one of which is involved in uric acid metabolism.
Silicon	Skeletal development, growth and maintenance of connective tissue
Vanadium	Growth, reproduction, fat metabolism
Arsenic	Growth, also some effect on blood formation, possibly haemoglobin production

As mentioned earlier, selenium is highly toxic in large doses and the available evidence suggests that the difference between the recommended allowance and the toxic dose may be quite small. Injudicious supplementation of foods is therefore particularly dangerous in this respect.

Cobalt

Cobalt is a component of vitamin B_{12} and this may be its only biological function in the dog and cat. Under laboratory conditions cobalt can replace zinc in a few zinc-containing enzyme systems but whether this is of biological importance is not known. In ruminants vitamin B_{12} can be synthesized by gut bacteria in the presence of cobalt in the upper intestine from which the vitamin is then absorbed. In non-ruminants like the dog and cat this synthesis may be of only limited use because it occurs mainly in the lower part of the intestine where absorption is minimal. It is likely that to be of significant nutritional value, cobalt must be ingested by the dog and cat, principally as vitamin B_{12}. With an adequate supply of the vitamin it is very doubtful whether any additional cobalt is required. Vitamin B_{12} will be discussed later in this chapter.

Other trace elements

A number of other trace elements have been demonstrated as necessary for normal health in mammals, although specific requirements have not been established for the dog and cat. These elements are listed in Table 2 with a brief summary of their functions. From work with other animals it appears that the amounts required in the diet are very low, usually well under 2 microgram (μg) per kcal, although silicon may be needed at a concentration of around 12 μg per kcal.

The likelihood of a deficiency of any of these nutrients in a normal diet is almost non-existent. Conversely, as with the majority of the trace elements these substances are all toxic if fed in large quantities, although the amounts which can be tolerated vary. Arsenic, vanadium, fluorine and molybdenum are the most toxic, whereas relatively large amounts of nickel and chromium can be ingested without adverse effects.

Vitamins

The vitamins may be conveniently divided into two sub-groups: fat soluble and water soluble. Apart from the obvious chemical difference the degree of storage in the body also differs, fat soluble vitamins being stored to a greater extent than the water soluble type. A regular daily supply is therefore less critical in the case of the fat soluble vitamins.

Fat soluble vitamins

VITAMIN A. The term vitamin A is now used to describe several biologically-active compounds but retinol is the substance of primary importance in mammalian physiology. In nature vitamin A is found to a large extent in the form of its precursors, the carotenoids, which are the yellow and orange pigments of most fruits and vegetables. Of these, β-carotene is the most important 'provitamin A' because it has the highest activity on a quantitative basis, consisting essentially of two vitamin A-type molecules linked together, which most animals can convert to two molecules of the active vitamin.

Here we find yet another important difference between the dog and cat. It has been shown that, unlike the dog, the cat is unable to convert β-carotene to vitamin A; cats therefore require a pre-formed dietary source of vitamin A of which the most common forms are derivatives of retinol (retinyl acetate and retinyl palmitate). The practical consequence of this peculiarity is that the cat must have at least some animal-derived material in its diet since *pre-formed* vitamin A compounds are not present in plants.

The best known function of vitamin A is in the physiological functions of vision. It is found in the retina combined with a specific protein called opsin. The compound is called rhodopsin (visual purple) and on exposure to light is split into opsin and a metabolite of retinol. It is the energy exchange in this process which produces nervous transmissions which are sent via the optic nerve to the brain and which result in visual sensations. Although the splitting of rhodopsin is reversible, a fresh supply of vitamin A is required to reform the visual pigment completely and so allow the process to continue. Vitamin A is involved in many other physiological functions one of the most important being the regulation of cell membranes; it is essential for the integrity of epithelial tissues and the normal growth of epithelial cells. Vitamin A is also involved in the growth of bones and teeth.

As might be expected, a deficiency of vitamin A has many far-reaching effects on the body and has been observed in many animals, including cats and dogs. The symptoms include xerophthalmia (excessive dryness of the eye), ataxia, conjunctivitis, opacity and ulceration of the cornea, skin lesions and disorders of the epithelial layers, e.g., the bronchial epithelium, respiratory tract, salivary glands and seminiferous tubules.

An *excess* of vitamin A is as harmful as a deficiency. A crippling bone disease with tenderness of the extremities associated with gingivitis and tooth loss has been described in cats given prolonged excessive doses of this vitamin either as vitamin A itself or by feeding large quantities of raw liver. Similar effects have been seen in dogs given large doses of vitamin A. Thus inclusion in the diet of foods containing large quantities of this vitamin, e.g. liver and the fish liver oils, must be very carefully controlled. Supplementation of an already adequate diet is not only unnecessary but potentially dangerous and should be avoided.

VITAMIN D. There are several compounds which have vitamin D activity but the two most important are called ergocalciferol (vitamin D_2) and cholecalciferol (vitamin D_3). Both of these forms are effective in dogs and cats as sources of vitamin D activity. There has been a large amount of research conducted on the metabolism of vitamin D in other mammals and it is now known that this vitamin undergoes a series of biochemical conversions in the kidney and liver before it becomes physiologically active. It is a dihydroxy derivative of the parent compound that is the most potent metabolite. Vitamin D is often called the 'bone vitamin' and its most clearly established function is to raise the plasma calcium and phosphorus levels to those required for the normal mineralization of bone. In the small intestine vitamin D stimulates the absorption of calcium and phosphorus and is also involved in the mobilization of calcium from bone to maintain a normal plasma calcium concentration. In fact the biochemical synthesis of the active vitamin D compound is triggered by a fall in plasma calcium. It is clear that the requirements for vitamin D are closely linked to the dietary concentrations of calcium and phosphorus and to the calcium/phosphorus ratio.

As vitamin D is involved in the absorption of calcium, it is most crucial during the growth and development of bone, i.e. in the young growing animal. A deficiency of this vitamin causes rickets. However there is evidence that dogs and cats, in common with other mammals, can form vitamin D_3 from lipid compounds in the skin in the presence of the ultra-violet component of sunlight and it is likely that adult animals need little if any *dietary* supply of this vitamin. Furthermore, Rivers et al. (1979) reported that cats are almost totally independent of a dietary source of vitamin D, even during growth and shielded from ultra-violet light, assuming they are fed a diet with adequate concentrations (and a correct ratio) of calcium and phosphorus. This appears to be because the cat mobilizes stores of vitamin D_3 which it acquires during suckling.

As with vitamin A, excessive amounts of vitamin D cause adverse effects in dogs and cats, notably extensive calcification of the soft tissues, lungs, kidneys and stomach. Deformations of the teeth and jaws can also occur and death can result if the intake of the vitamin is particularly high. Supplementation of vitamin D is therefore potentially hazardous, and for the cat the requirement may be so low that any reasonable diet is bound to supply adequate quantities.

VITAMIN E. The function which was first ascribed to this vitamin was that of preventing foetal resorption in animals that had been fed a diet containing rancid lard. The chemical name for this vitamin (tocopherol) is derived from the Greek word 'to bring forth offspring'. However in recent years studies on Vitamin E have revealed much more of its role in the body although complete details of its function remain obscure.

It acts as an anti-oxidant and is important in maintaining the stability of cell membranes; in this its function is closely linked to that of the trace element selenium which was discussed earlier. The requirement for vitamin E also depends on the level of polyunsaturated fatty acids (PUFA) in the diet. Increasing PUFA increases the vitamin E requirement and this effect has been shown in many animals, including the dog and cat. It is difficult therefore to be precise about recommendations for dietary vitamin E; the levels stated in Appendices I and II are based on normal dietary concentrations of selenium and PUFA. Rancid fats should be avoided because they are particularly destructive to this vitamin.

Deficiency of vitamin E under experimental conditions presents a more bewildering range of physical abnormalities than is encountered with any other vitamin. These effects may be divided into four main areas: the muscle, reproductive, nervous and vascular systems. In dogs, a deficiency has been associated with one or more of these effects including skeletal muscle dystrophy, degeneration of the germinal epithelium of the testes, and failure of gestation. Vitamin E deficiency in dogs has also been linked with impairment of the immune response. In cats inflammatory changes in body fat (steatitis — 'yellow fat disease') occurs when low levels of vitamin E are fed in the presence of PUFA.

There is only very limited information on the effects of high vitamin E intakes in the dog and cat. No deleterious effects were reported when about 10 times the recommended level was fed to weaned Beagle puppies for 15 weeks. However in other species some adverse reactions on thyroid activity and blood clotting have been noted with high vitamin E intakes. The latter probably occurs via inhibition of vitamin K activity (see next section). Therefore high levels of this nutrient must be considered potentially harmful, although it is far from being as dangerous as vitamins A and D in excess.

VITAMIN K. Vitamin K describes a group of compounds, the quinone derivatives, which regulate the formation of several factors involved in the blood clotting mechanism. A requirement for vitamin K has been demonstrated in the dog and it is unlikely that the cat is any different in this respect. Nevertheless the requirement in dogs was demonstrated under experimental conditions where the animals were made vitamin K deficient by the use of anti-coagulant drugs (such as the coumarin compounds) which antagonize the action of this nutrient. In normal healthy animals a vitamin K deficiency is very rare because dogs and cats, like other mammals, obtain most if not all of their daily requirement from bacterial synthesis in the intestine. It is only under abnormal conditions such as depression of bacterial synthesis (for example by drug treatment) and interference with the absorption or utilization of vitamin K, that a dietary supply will be necessary. A diet containing only 60 μg/kg dry matter (about 0.017 μg per kcal) was fed to adult male Beagles and cats for 40 weeks with no signs of deficiency although the same diet resulted in haemorrhages when fed to rats. A concentration of 0.02 μg per kcal has been suggested as a minimum requirement for cats, although this is probably necessary only when bacterial synthesis has been suppressed or there are antivitamin K components in the diet. Very large intakes of vitamin K produce anaemia and other blood abnormalities in young animals but it does not appear to be particularly toxic.

Water soluble vitamins

The water soluble vitamins of importance in dog and cat nutrition are all members of the B complex and nearly all are involved with the utilization of foods and the production or interconversion of energy in the body. In these processes, the B vitamins are used by the animal to form coenzymes (sometimes also called cofactors). These are relatively small organic molecules, associated with larger enzyme molecules, which are necessary for the enzymes to catalyse biochemical reactions effectively. The coenzymes often act by combining with and then releasing molecules or fragments of molecules, rather like a biochemical 'relay station'. Sometimes minerals and trace elements are also involved in these reactions as has been discussed earlier in the text.

The B vitamins are now usually known by their chemical names, rather than by a letter/number combination, but this alternative nomenclature will be mentioned for vitamins where it is still in common use.

THIAMIN (ANEURIN, VITAMIN B_1). Thiamin is a sulphur-containing compound which participates as a coenzyme in the form of its pyrophosphate (TPP), sometimes referred to as cocarboxylase. TPP is involved in several key conversions in carbohydrate metabolism and the thiamin requirement is dependent on the carbohydrate content of the diet. A high fat, low carbohydrate diet will spare the need for thiamin as less of this vitamin is required for fat metabolism than in carbohydrate utilization.

Thiamin deficiency has been described in dogs and cats. Its primary effect is a 'biochemical lesion' resulting in impaired carbohydrate metabolism with abnormal accumulation of the intermediate compounds of the metabolic pathway. The deficiency expresses itself clinically as anorexia, neurological disorders (especially of the postural mechanisms) followed ultimately by weakness, heart failure and death. In man, thiamin deficiency is known as beri-beri. Thiamin is a particularly important vitamin from the aspect of dietary formulation because it is progressively destroyed by cooking and may also be inactivated by naturally occurring substances called thiaminases which are found in a number of foods, particularly raw fish. Thiaminases are themselves inactivated by heat so the maintenance of an adequate thiamin intake must take all of these various factors into consideration. For commercially-prepared foods, the normal practice is to supplement with a large enough quantity before processing so that even if particularly serious losses occur, the amount remaining in the finished product will still meet or exceed the dietary recommendations.

Like the other water soluble vitamins, thiamin is of low toxicity. Although intravenous *injection* of thiamin in dogs produces death through depression of the respiratory centre, the *oral* intake needed to cause the same effect is some 40 times the intravenous dose and represents a level many thousands of times the recommended dietary concentration.

RIBOFLAVIN (VITAMIN B_2). Riboflavin is a yellow crystalline compound which shows a characteristic yellow-green fluorescence when dissolved in water. Riboflavin is a constituent of two coenzymes, riboflavin 5-phosphate and a more complex chemical called flavin adenine dinucleotide. These coenzymes are essential in a number of oxidative enzyme systems. Cellular growth cannot occur in the absence of riboflavin.

Riboflavin requirements have been investigated in dogs and cats and a deficiency is associated with eye lesions, skin disorders and testicular hypoplasia. There is some evidence that part of the requirement for riboflavin can be met by bacterial synthesis in the intestine and that this is favoured by a high-carbohydrate, low-fat diet. However, the daily needs for the vitamin are certainly greater than any possible contribution by this route so a regular dietary intake is necessary.

PANTOTHENIC ACID. This substance is a constituent of coenzyme A which is an essential component of enzyme reactions in carbohydrate, fat and amino-acid metabolism. A need for pantothenic acid has been demonstrated in dogs and cats. There are many deficiency signs, including depression or failure of growth, development of fatty liver and gastrointestinal disturbances including ulcers. In dogs, but not cats, alopecia has also been observed. These deficiency signs were produced using semi-purified diets. Under normal circumstances, using a mixture of foods, a deficiency of pantothenic acid is extremely unlikely as it is very widespread in animal and plant tissues, as implied by its name which means 'derived from everywhere'.

NIACIN (NICOTINIC ACID). Niacin is rapidly converted in the body to the physiologically active derivative nicotinamide (niacinamide). Nicotinamide is a component of two very important coenzymes, the nicotinamide adenine dinucleotides, which are required for oxidation–reduction reactions necessary for the utilization of all the major nutrients. In mammalian species, including the dog, the requirement for niacin is influenced by the dietary level of the amino-acid tryptophan, which can be converted to the vitamin. In cats this conversion does not occur but, unlike the other differences between the dog and cat, this is not due to lack of an enzyme. It occurs because the reaction sequence for the breakdown of tryptophan can go one of two ways and in the cat the enzyme responsible for the alternative 'non-niacin' pathway has a very high activity and effectively abstracts the tryptophan metabolites from niacin synthesis. This alternative pathway eventually breaks down the metabolites to supply energy, similar to the utilization of carbohydrates.

Niacin deficiency has been described in dogs and cats and is accompanied by inflammation and ulceration of the oral cavity with thick, blood-stained saliva drooling from the mouth, and foul breath. The deficiency syndrome is referred to as blacktongue in the dog and pellagra in man. Niacin is sometimes called the pellagra-preventing vitamin or PP factor. Large doses of niacin (but not nicotinamide) produce a flushing reaction in many animals including dogs. Thus if large therapeutic quantities of this vitamin need to be adminstered, it is preferable to use the amide form. Nevertheless, neither of the two forms of this vitamin could be described as highly toxic.

PYRIDOXINE (VITAMIN B_6). There exist three related compounds under this heading with essentially equally effective activity: pyridoxine, pyridoxal and pyridoxamine. All three occur naturally and are interconvertible during normal metabolic processes. The biologically active compound is pyridoxal and the coenzyme form is pyridoxal 5-phosphate, which is involved in a large number and variety of enzyme systems almost entirely associated with nitrogen and amino-acid metabolism. In fact pyridoxal is considered essential for practically all enzymic interconversions and non-oxidative degradations of amino-acids. Some of these reactions have already been discussed in relation to other nutrients; for example, the synthesis of niacin from tryptophan involves this vitamin. As might be expected, a high protein diet exacerbates vitamin B_6 deficiency, an effect which is comparable with the effect of high carbohydrate diets on thiamin deficiency.

Pyridoxine has been shown to be required by dogs and cats; a deficiency results in weight loss and a type of anaemia. In cats, irreversible kidney damage can also occur with tubular deposits of calcium oxalate crystals (pyridoxine is required in the conversion of oxalate to glycine). Dermatitis and alopecia have occasionally been reported in pyridoxine deficiency in dogs. Like the other water soluble vitamins, pyridoxine and its derivatives are not considered highly toxic.

BIOTIN. Like other B vitamins, biotin is presumed to function as a coenzyme and is necessary for certain reactions involving the metabolism of the carboxyl (CO_2) group which is initially bound to the biotin before transfer to an 'acceptor' molecule. In biotin deficiency there is a reduction in amino-acid incorporation into proteins, apparently due to a fall in the synthesis of dicarboxylic acids. Impairment of glucose utilization and fatty acid synthesis have also been reported. In the early stages of deficiency the principal clinical sign seems to be a scaly dermatitis. Although these effects were initially investigated using other animals, it is now known that biotin is required by dogs and cats, and similar deficiency signs have been described. However, it is very difficult to produce biotin deficiency with a normal diet because most, if not all of the daily requirement can be met via synthesis by the gut bacteria. Deficiency signs have been produced in the dog and cat only when antibiotics were given to suppress bacterial action, and large quantities of whole egg white were included in the diet. Egg white contains a protein called avidin which forms a stable and biologically inactive complex with biotin. Avidin will also 'neutralize' biotin in food as well as that produced by bacteria. Avidin itself is relatively heat sensitive, so if eggs constitute a significant proportion of the diet they should be fed cooked rather than raw. It is also important to realize that antibiotic drugs can increase the requirement for vitamins like biotin (see also vitamin K and folic acid) because they destroy the intestinal bacteria responsible for their manufacture. Nevertheless the likelihood of a naturally occurring biotin deficiency is remote.

FOLIC ACID (PTEROYLGLUTAMIC ACID, FOLACIN). Folic acid is usually found in nature in the form of conjugates with the amino-acid, glutamic acid. The biologically active coenzyme is the tetrahydro derivative, often abbreviated as THFA or FH_4. There are several other forms of THFA with coenzyme activity all of which are usually grouped under the generic name of the folates or folate coenzymes. The folates are involved in the transfer of single carbon groups (e.g. methyl and formyl) which are important in several ways. Perhaps the most significant reactions are those necessary for the synthesis of thymidine, an essential component of the nucleic acid DNA. Lack of an adequate supply of DNA prevents normal maturation of primordial red blood cells in bone marrow and the typical signs of folic acid deficiency are therefore anaemia and leukopenia. Folic acid deficiency has been described in dogs and cats but usually only when semi-purified diets were fed in the presence of antibiotics. It is likely that most of the daily requirement for folate is met by bacterial synthesis in the intestine.

VITAMIN B_{12}. This vitamin is unique in being the first cobalt-containing substance shown to be essential for life and is the only vitamin that contains a trace element. Vitamin B_{12} is also known as cobalamin but is usually isolated in combination with a cyanide group linked to the cobalt atom. This form is known as cyanocobalamin and is sometimes used as a synonym for vitamin B_{12} itself. The active coenzyme form is yet another derivative where a new chemical group replaces cyanide in the parent molecule. Like the folates vitamin B_{12} is involved in the transfer of single carbon fragments and its function is closely linked to that of folic acid itself.

Vitamin B_{12} is also involved in fat and carbohydrate metabolism and in the synthesis of myelin, a constituent of nerve tissue. The typical signs of a B_{12} deficiency in many ways resemble those of folate deficiency but characteristically also involve neurological impairment as a result of inadequate production of myelin. Vitamin B_{12} is only poorly absorbed from ingested food unless a protein, 'intrinsic factor', is present in the intestine. This factor presumably facilitates transfer of the vitamin across the mucosal membrane. Failure to absorb B_{12} due to lack of intrinisic factor, results in pernicious anaemia with neurological degeneration.

These effects have been described in other animals including man, but less information is available for the dog and cat. The vitamin has been shown to be needed by these two species but a quantitative requirement has not been determined in detail. The stated requirements are based on some research conducted in dogs and cats, and on data from other mammals.

CHOLINE. Choline is a component of the phospholipids which are essential components of cell membranes, it is the precursor of acetylcholine, one of the body's neurotransmitter chemicals. It is an important methyl donor, that is it supplies single carbon fragments for metabolic conversions, the significance of which have already been discussed in the previous sections dealing with folic acid and vitamin B_{12}. A deficiency of choline causes several abnormalities including kidney and liver dysfunction which in the dog and cat are usually manifested as fatty infiltration of the liver. The precise mechanism for this is not known but may be linked to inadequate biosynthesis of specific types of phospholipids, leading to impaired rates of lipid transport.

The requirement for choline in the diet can be modified by a number of factors, in particular the dietary concentration of methionine. Since methionine can also act as a methyl donor in intermediary metabolism, an increased dietary supply of one tends to spare the need for the other. Some work with cats has shown that methionine can completely replace the dietary need for choline if supplied in adequate amounts (Anderson *et al.* 1979). In view of the sparing effect of methionine and the widespread distribution of choline in plant and animal materials it is most unlikely that a dog or cat will become choline deficient under normal circumstances.

ASCORBIC ACID (VITAMIN C). Unlike man, dogs and cats do not need a dietary supply of this vitamin because they are able to synthesize it from glucose. Nevertheless some researchers have claimed that a number of diseases of the dog can be ameliorated by ascorbic acid. Furthermore skeletal diseases such as hypertrophic osteodystrophy, hip dysplasia and a number of others, particularly those common in the large and giant breeds, have been said by some to resemble ascorbic acid deficiency (scurvy). However other research groups have consistently failed to show any benefits of vitamin C in either alleviating or preventing these diseases. It is possible that some individual dogs could have a reduced capacity to synthesize this vitamin but on the available evidence it appears that there is no general dietary requirement for this nutrient by dogs and cats.

DIETARY RECOMMENDATIONS

Early in this chapter much emphasis was placed on the important difference between nutrient requirements and recommended levels. The NRC values (Appendices IV and V) are a useful and concise summary of the current knowledge on the minimum requirements of nutrients in foods for dogs and cats. However, it is clear from the points made in this and other chapters in the book that recommended contents of nutrients in a ‘practical’ diet must incorporate, as far as possible, allowances for nutrient interactions, nutrient availability, animal variability and so on. Nutritionists at the Waltham Centre for Pet Nutrition have developed a list of recommended nutrient concentrations for diets designed as complete and balanced for all life stages of dogs and cats and these are summarized in Table 3. These values are based to a large extent on NRC data and, as with all aspects of nutritional recommendations, incorporate certain assumptions as to nutrient availability, protein amino-acid balance and type of ingredients. The values are stated in terms of metabolizable energy (ME) so that they are applicable to any food, regardless of its moisture content. The ME content chosen is 400 kcal. Although this may seem rather arbitrary, it is based on the fact that most commercial dog and cat foods contain approximately 400 kcal ME per 100g of dry matter. Thus the values in Table 3 can also be approximated to a concentration per 100 g of dry matter of the diet.

These figures are *not* a guarantee of nutritional quality but are good guidelines to produce a correctly balanced diet. The guarantee of nutritional adequacy comes only from animal feeding studies.

TABLE 3

Recommended minimum nutrient levels in dog and cat diets (from WCPN)

Nutrient		Dogs	Cats	Notes
Protein	g	22	28	a
Fat	g	5.5	9	b
Linoleic & arachidonic acids	g	1.1	1.0	b
Arachidonic acid alone	g	—	0.02	b
Calcium	g	1.1	1.0	
Phosphorus	g	0.9	0.8	
Ca/P ratio		0.8:1.0	0.8:1.0	
Sodium	g	0.2	0.2	
Potassium	g	0.5	0.4	
Magnesium	g	0.04	0.05	
Iron	mg	8.0	10	c
Copper	mg	0.7	0.5	c
Manganese	mg	0.5	1.0	c
Zinc	mg	5.0	4.0	c
Iodine	mg	0.15	0.1	
Selenium	μg	10	10	
Vitamin A	IU	500	550	
Vitamin D	IU	50	100	
Vitamin E	mg	5.0	8.0	b
Vitamin K	μg	8.0	8.0	d
Thiamin	mg	0.1	0.5	
Riboflavin	mg	0.25	0.5	
Pantothenic acid	mg	1.1	1.0	
Niacin	mg	1.2	4.5	
Pyridoxine	mg	0.12	0.4	
Folic Acid	μg	22	100	d
Vitamin B_{12}	μg	2.7	2.0	
Choline	mg	125	200	
Biotin	mg	—	—	d
Taurine	mg	—	100	c

All values expressed per 400 kcal ME.

(a) Protein levels assume a balanced amino-acid profile and satisfactory digestibility.

(b) Fat content stated only as a guide. Key nutrients are the EFA linoleic and arachidonic acids. With high levels of EFA, vitamin E will need to be increased.

(c) Figures assume high availability. Particularly important to ensure that this is the case with these nutrients.

(d) A metabolic requirement for these nutrients has not been demonstrated when natural ingredients were fed. This is because intestinal bacterial synthesis can meet the needs of the animal. Supplementation may be necessary if antibacterial or antivitamin compounds are being administered or present in the diet.

SUMMARY

When the first edition of this book was published in 1982 it was predicted that the area of expansion in nutritional knowledge would be nutrient interaction. This is even more true today and it is again the first of three themes that are worthy of highlighting.

No nutrient functions in isolation. The various interactions that, for example, affect zinc availability are an excellent illustration of the care that must be taken in transposing experimental data into practical feeding and the formulation of pet foods.

The second message is that all nutrient requirements are finite: enough is enough. No magical, health promoting or life-prolonging properties result from an excessive intake of any nutrient. Unjustified supplementation is not only unnecessary but can be downright dangerous, particularly in the case of the trace elements and fat soluble vitamins. Furthermore, although the discussion of requirements has always been in terms of a daily intake of nutrients, this should not give the impression that dogs and cats must receive the precise recommended intake every day without fail. Even for the water soluble vitamins which are not stored to any extent in the body, slight fluctuations can be tolerated by the healthy animal without any ill-effects as long as the *average* amount ingested over say, two or three days is sufficient. It is only in extreme cases like very low concentrations (or total absence) of a nutrient or an unusual metabolic effect (such as arginine deficiency in the cat) that the animal is likely to come to harm.

Lastly, the most interesting aspect of dog and cat nutrition is probably the differences between their requirements, in particular the atypical metabolism of the cat. The overall theme is that the cat is dependent on a supply of at least some animal-derived materials in its diet and must be regarded as an obligate carnivore. Why should it not possess fully active enzyme systems responsible for the production of taurine, EFA and vitamin A, and the conservation of protein? Is it possible that during the course of evolution these functions have been lost because of the ability of the cat family to catch prey and live on what is an almost entirely animal-based diet? Alternatively, were the early mammals obligate carnivores and the cat family represents an early branch of the evolutionary tree with such an efficient predatory lifestyle that it was subjected to little or no environmental pressure to develop mechanisms to utilize plant raw materials? Perhaps an investigation of the nutritional requirements of a primitive placental mammal like the hedgehog would reveal some interesting facts to contribute to the discussion!

Whatever the reasons for these differences between the dog and cat it must always be remembered by those involved in any aspect of pet feeding that, nutritionally and biochemically, a cat is not just a small, highly agile dog that climbs trees.

BIBLIOGRAPHY

Anderson, P. A., Baker, D. H., Sherry, P. A. and Corbin J. E. (1979) Choline–methionine interrelationship in feline nutrition. *J. Anim. Sci.* **49,** 522–527.

Blaza, S. E., Booles, D. and Burger, I. H. (1988) Is carbohydrate essential for pregnancy and lactation in dogs? Proceedings of the Waltham Symposium No. 7 (1985) Cambridge University Press.

Burger, I. H., Blaza, S. E., Kendall, P. T. and Smith P. M. (1984) The protein requirement of adult cats for maintenance. *Feline Pract.* **14,** 8–14.

Chausow, D. G. and Czarnecki-Maulden, G. L. (1987) Estimation of the dietary iron requirement for the weanling puppy and kitten. *J. Nutr.* **117,** 928–932.

Hayes, K. C., Carey, R. E. and Schmidt, S. Y. (1975) Retinal degeneration associated with taurine deficiency in the cat. *Science* **188,** 949–951.

Kronfeld, D. S., Hammel, E. P., Ramberg Jnr., C. F. and Dunlop Jnr. H. L. (1977) Haematological and metabolic responses to training in racing sled dogs fed diets containing medium, low or zero carbohydrate. *Am. J. Clin. Nutr.* **30,** 419–430.

MacDonald, M. L., Anderson, B. C., Rogers, Q. R., Buffington, C. A. and Morris, J. G. (1984a) Essential fatty acid requirements of cats: pathology of essential fatty acid deficiency. *Am. J. Vet. Res.* **45,** 1310–1317.

MacDonald, M. L., Rogers, Q. R. and Morris, J. G. (1984b) Nutrition of the domestic cat, a mammalian carnivore. *Ann. Rev. Nutr.* **4,** 521–562.

Morris, J. G. and Rogers, Q. R. (1978) Arginine: an essential amino-acid for the cat. *J. Nutr.* **108,** 1944–1953.

NRC (1985) Nutrient requirements of dogs, National Research Council, National Academy of Sciences, Washington D.C.

NRC (1986) Nutrient requirements of cats, National Research Council, National Academy of Sciences, Washington D.C.

Pion, P. D., Kittleson, M. D., Rogers, Q. R. and Morris, J. G. (1987) Myocardial failure in cats associated with low plasma taurine: a reversible cardiomyopathy. *Science* **237,** 764–768.

Rivers, J. P. W. (1982) Essential fatty acids in cats. *J. Small Anim. Pract.* **23,** 563–576.

Rivers, J. P. W., Frankel, T. L., Juttla, S. and Hay, A. W. M. (1979) Vitamin D in the nutrition of the cat. *Proc. Nutr. Soc.* **38,** 36A.

Rogers, Q. R. and Morris, J. G. (1982) Do cats really need more protein? *J. Small Anim. Pract.* **23,** 521–532.

Romsos, D. R., Plamer, H. J., Muiruri, K. L. and Bennink, M. R. (1981) Influence of a low carbohydrate diet on performance of pregnant and lactating dogs. *J. Nutr.* **111,** 678–689.

Sinclair, A. J., Slattery, W., McLean, J. G. and Monger, E. A. (1981) Essential fatty acid deficiency and evidence for arachidonate synthesis in the cat. *Br. J. Nutr.* **46,** 93–96.

Sturman, J. A., Gargano, A. D., Messing, J. M. and Imaki, H. (1986) Feline maternal taurine deficiency: effect on mother and offspring. *J. Nutr.* **116,** 655–667.

Thornburg, L. P., Ebinger, W. L., McAllister, D. and Hoekema, D. J. (1985a) Copper toxicosis in dogs, Part I: copper-associated liver disease in Bedlington Terriers. *Canine Pract.* **12,** 41–45.

Thornburg, L. P., Dennis, G. L., Olwin, D. B., McLaughlin, C. D. and Gulbas, N. K. (1985b) Copper toxicosis in dogs, Part 2: the pathogenesis of copper-associated liver disease in dogs. *Canine Pract.* **12,** 33–38.

Van den Broek, A. H. M. and Thoday, K. L. (1986) Skin diseases in dogs associated with zinc deficiency: a report of five cases. *J. Small Anim. Pract.* **27,** 313–323.

CHAPTER 3

Digestion, Absorption and Dietary Balance

IVAN H. BURGER AND SANDRA E. BLAZA

In the previous chapter the various aspects of nutrients and their requirements by dogs and cats were discussed. In this chapter the ways in which dietary constituents are metabolized (i.e. broken down and utilized) by the animal will be considered.

The body is not a sealed unit with a fixed composition, but is in a continual state of change; meals are eaten and nutrients absorbed and utilized or excreted. The maintenance of the level of any single nutrient, or the energy or water content of the body, can be thought of in terms of 'balance'. For many of the nutrients required by mammals, such as proteins, fats, minerals and vitamins, balance is achieved largely by the control of output. The prime requirement for mammals is to satisfy their energy needs. The level of intake of particular nutrients is therefore dependent upon their concentration in relation to food energy (see Chapter 2). Provided that the necesary minimum intake is met, anything surplus to requirements can be lost in the faeces, or having been absorbed can be converted to other useful substances (in the liver) and used or excreted via the kidneys in the urine.

The regulation of energy balance and fluid balance is rather more complex than that of individual nutrients, as both involve mechanisms which govern intake as well as output. The emphasis is different for energy and fluid, so these will be covered separately. As digestion and absorption are central to the maintenance of both, these will be considered later.

ENERGY BALANCE

An animal is said to be in energy balance when its expenditure of energy is equal to its intake, with the result that the level of energy stored in the body does not change. The following equation is derived from the Law of Conservation of Energy

$$\text{Energy stored} = \text{energy intake} - \text{energy expenditure}$$

In the adult dog or cat, energy is stored predominantly as fat with some

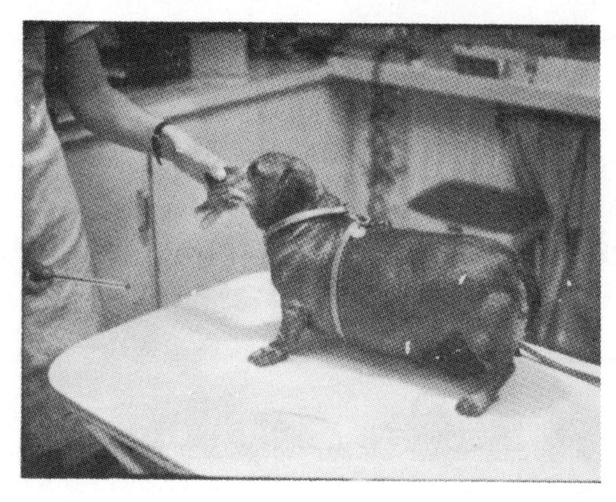

FIG. 1. Obesity.

increase in lean tissue (fat-free mass). In a growing or pregnant animal, the emphasis shifts to the accumulation of lean tissue. Fat is stored as adipose tissue; these deposits are easily observed in an obese animal (Fig.1). It is also possible to decrease energy stores, by reducing energy intake until it is less than energy output. This reverses the equation.

Expenditure – intake = loss of energy from stores

Under these conditions of negative energy balance, the body has to catabolize (break down) its own tissues to meet its energy needs; as the stores are gradually depleted, the animal becomes thinner and bodyweight decreases.

Energy balance is achieved by the exact matching of input and output over long periods. A very small imbalance maintained for a long time will cause obesity (if the net difference is positive) or wasting (if the difference is negative). For example, imagine a Labrador Retriever which has a daily energy expenditure of 1700 kcal but an intake of 1800 kcal. The imbalance is only 100 kcal per day (approximately 1 oz, 30g, of dog biscuit), but if maintained could result in an increase of 2–3 kg (4–6lb) in bodyweight over a year. If this rate of gain continues for 2 or 3 years, even allowing for some compensation of energy output (see below) the dog will become very overweight and subject to all the problems of obesity.

Until recently it has been assumed that energy balance relied only on the exact regulation of intake and that any imbalance represented a failure to control intake appropriately. It now seems probable that both intake and output are important in the maintenance of energy balance. Intake is capable of precise control in contrast to expenditure which appears to act as a crude buffer, opposing any change to the energy content of the body.

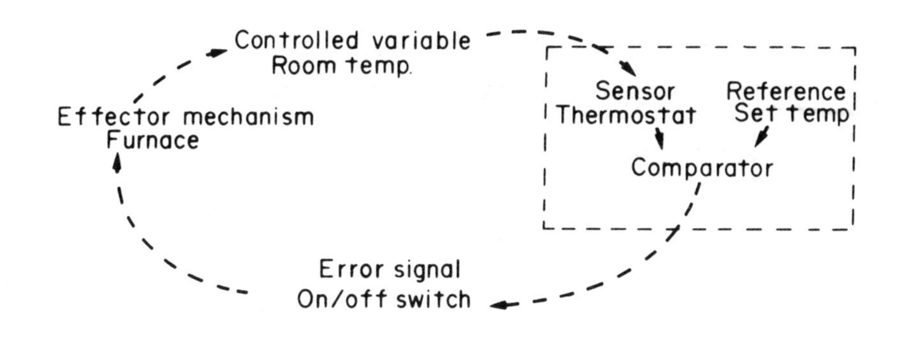

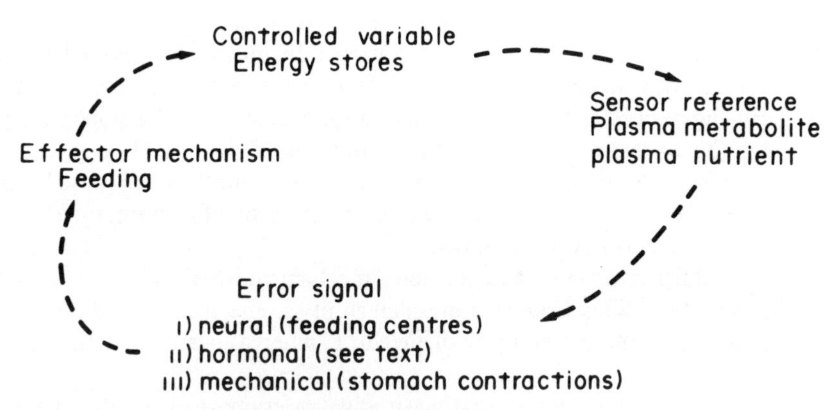

FIG. 2. Negative feedback.

Regulation of Energy Balance Through Intake

Although very palatable foods may disturb the control of intake, many animals fed bland foods regulate energy intake very precisely. Even when the same food is given in dilute form (using water or some non-digestible material to change the energy density) compensation usually occurs rapidly and completely to a new appropriate level of food intake. There are many theories which attempt to account for this and most fit the negative feedback mechanism described in Fig. 2, to a greater or lesser degree.

The principle of negative feedback is a very simple one which can be demonstrated in many forms. For example, the control of blood pressure is dependent on negative feedback, as are many of the body's control mechanisms. It is found frequently in engineering as the example in Fig. 2 shows.

In its most straightforward form, negative feedback is a system where any change in the equilibrium of the system elicits a signal, provoking a response to oppose the initial change, and correct the error. In the example the room thermostat is the *sensor* which detects any change in ambient temperature. The discrepancy between the room temperature and a set reference temperature is noted by the *comparator* which signals to the boiler or *effector*, switching it on or off. The heat output from the boiler restores the room temperature, which can be thought of as the *controlled variable* and causes the comparator to cease signalling to the boiler.

In this model of the regulation of energy balance in Fig. 2, the controlled variable is the size of the energy stores. There are several feedback elements which may signal change, such as plasma nutrient and metabolite levels. Any discrepancies against set reference points indicate change in energy stores and stimulate neural and hormonal activity which may initiate or inhibit feeding.

The neural response involves 'feeding centres' in the brain which are not discrete 'hunger' and 'satiety' centres as once thought, but bundles of neurons covering several areas. Stimulation of these by electrodes can cause satiated animals to eat or prevent hungry animals from eating.

The hormonal response is more complex. Insulin stimulates feeding but it is not known whether this is a direct effect on the central nervous system or by causing peripheral hypoglycaemia (low glucose levels in the blood). Glycogen has the opposite effect to insulin, inhibiting feeding, as do oestrogens and luteinizing hormone (female reproductive hormones). As all these hormones have roles other than the stimulation or inhibition of feeding, they cannot be the sole agents governing intake.

In addition to neural and hormonal mechanisms there are other, more direct stimuli to feeding. Contractions of an empty stomach are thought to cause the sensation of hunger and provoke feeding, whereas gastric distension inhibits eating.

This model is useful in that it gives a framework to the theories on regulation of energy balance through intake, and it also allows testing of the theories. However, much of it is speculation and allowance has to be made for some contribution of energy expenditure to the regulation of energy balance. In most cases regulation is successful. Disturbances occur when highly palatable titbits are offered and the appetite for a particular food drowns out physiological satiety signals: this seems to affect dogs rather more than cats which suffer much less often from obesity.

Energy Expenditure

Since all the energy expended by the body can be measured as heat, energy output is often referred to in terms of 'heat loss', 'heat production', or in units usually associated with the measurement of heat such as watts, joules and calories. These units may also apply to the measurement of energy intake.

Energy expenditure can be divided into two parts, basal metabolic rate (BMR) and thermogenesis. BMR is the amount of energy required to keep the body 'ticking over', that is it represents the energy needed to meet the cost of essential work done by the cells and organs. This includes such processes as respiration, circulation and kidney function. Many factors determine BMR in any individual, including bodyweight and composition, age and hormonal status (particularly the thyroid hormones). As these factors change, so does the rate of basal metabolism, although such changes tend to occur slowly over long periods.

Additional energy expenditure comes under the collective title of thermogenesis. This can be the cost of digesting, absorbing and utilizing nutrients (sometimes called the 'thermic effect of food', or 'dietary induced thermogenesis'), of muscular work or exercise, of stress or of the maintenance of body temperature in a cold environment. The intake of certain drugs or hormones can also cause thermogenesis. Thermogenesis is simply any increase in metabolic rate over the basal level. In contrast to BMR the degree of thermogenesis can vary widely and quickly, and may cause large daily variations in energy output. Of the two components of total energy expenditure, thermogenesis is the part capable of rapid adaptive response to changes in the internal or external environment.

The Regulation of Energy Balance Through Expenditure

When food is restricted for a long time, BMR decreases in a two-stage response. The first sharp drop is seen within days of the initial reduction in food intake and is thought to be a depression in the metabolism of individual cells, achieved through the thyroid and adrenal hormones which regulate metabolic rate. If refeeding occurs during this phase, metabolic rate recovers its initial level very quickly. The second phase takes longer to appear: this is a very gradual decline following the loss of body tissue, particularly lean tissue which is metabolically very active. Once this phase has been reached, BMR cannot be restored until the tissue has been replaced.

This decrease is a very simple form of regulation and will reduce any loss of bodyweight as a result of food restriction, although it cannot prevent it. There are important implications for the maintenance of bodyweight following a slimming regime; the level of food intake required to maintain energy balance will be less than required at the previous weight.

Similarly prolonged overfeeding results in an elevated metabolic rate which restricts the increase in energy stores. This increase in energy expenditure is partly attributable to the thermic effect of the food and to the cost of maintaining extra body tissue. However, these do not completely account for the extra expenditure, and there is still controversy over the source of the additional heat production.

There is, therefore, a rudimentary regulation of energy output which opposes any change in the status quo. Although energy output cannot completely prevent any change, it can limit its extent and this contribution should not be ignored.

WATER BALANCE

Water is often neglected as a nutritional requirement because of its ready availability in most temperate climates. However the requirement for water is at least as important as that for other nutrients; life may continue for weeks in the complete absence of food but only for days or even hours when water is not available.

Water fulfils many roles within the body. It is an excellent solvent, and this property makes possible all the complex chemistry of cellular metabolism. As the principal constituent of blood, water provides a vital transport medium, taking oxygen and nutrients to the tissues and removing carbon dioxide and metabolites. Blood also carries antibodies and white cells which protect the body from disease.

Water contributes to temperature regulation in several different ways. Firstly, the blood transports heat away from working organs and tissues, thereby preventing dangerous temperature increases. Then, by redirection of some of the blood through superficial veins, heat can be transferred to the skin and lost to the environment by radiation, convection and conduction. Heat loss may be further increased by the evaporation of water from the skin.

Water is also essential for digestion. Hydrolysis, the splitting of compounds by water, is the means by which digestion occurs. Digestive enzymes are secreted in solution, the better to disperse amongst the foodstuffs. Even the elimination of toxic metabolites via the kidney requires water as a medium. These represent only a few of the many functions of water.

There are several different fluid compartments in the body, which can be grouped together as intra- or extra-cellular fluid (ICF and ECF). ICF represents approximately 50% of the animal's total bodyweight and includes the water inside all cells from red blood cells to the neurons in the spinal cord. ECF is found bathing the tissues in between the cells, and in the blood and lymph. Movement of fluid between these compartments is continuous, different concentrations of electrolytes being maintained by the activity of cell membranes.

Water Output

Water leaves the body by several routes. In the normal healthy dog and cat these include losses in expired air, in faeces, in urine and rarely also in sweat.

These pathways will be discussed separately. In sick animals water loss may be increased markedly through haemorrhage (bleeding), vomiting and diarrhoea. Lactation is another instance of increased loss.

Faeces

The water content of faeces is usually very low compared with the enormous volumes of fluid secreted into the digestive tract, with enzymes, mucus and various electrolytes. The intestines have very efficient mechanisms for water reabsorption and it is only when these are disturbed and faeces evacuated as diarrhoea that this route makes a significant contribution to water loss.

Evaporative losses

The uptake of oxygen from inspired air is made possible by close association between the epithelium of the lung and an extensive capillary network. However, this also facilitates the transfer of water by diffusion and evaporation into the cavity of the lung, and the water is then lost in expired air. This 'respiratory water loss' is unavoidable. In hot weather evaporation is an important temperature regulating mechanism because of the body heat used to vaporize the water. This is why dogs pant and hang out their tongues, and why cats cover their coats with saliva by repeated licking. In extreme conditions there may also be some slight evaporation through the foot pads. Although these mechanisms aid temperature control, they may increase water loss..

Urine

The kidney is the only organ in the body which can control water loss. In addition to this it also regulates acid–base balance and the concentration of many electrolytes. In common with other mammals, dogs and cats have two kidneys situated in the abdominal cavity, one either side of, but ventral to (below, or in front of) the spinal column (Fig. 3). The blood supply is provided by the renal artery and vein.

The kidney consists of a network of thousands of tubules (Fig. 4). Each tubule has a blind end, or 'glomerular capsule' which envelops a knot of capillary blood vessels known as the glomerulus. There is a wide difference in pressure between the capillary and the capsule and this differential causes continuous movement of small molecules and fluid into the capsule from the capillary. Large molecules, such as proteins and the various blood cells, cannot pass into the tubule unless there has been damage to the glomerular or tubular walls. Indeed, one of the indications of kidney damage is the finding of proteins in the urine. In healthy animals therefore, the fluid which enters the tubule is an 'ultrafiltrate' of blood, and the rate of entry depends upon the difference in pressures in the two systems.

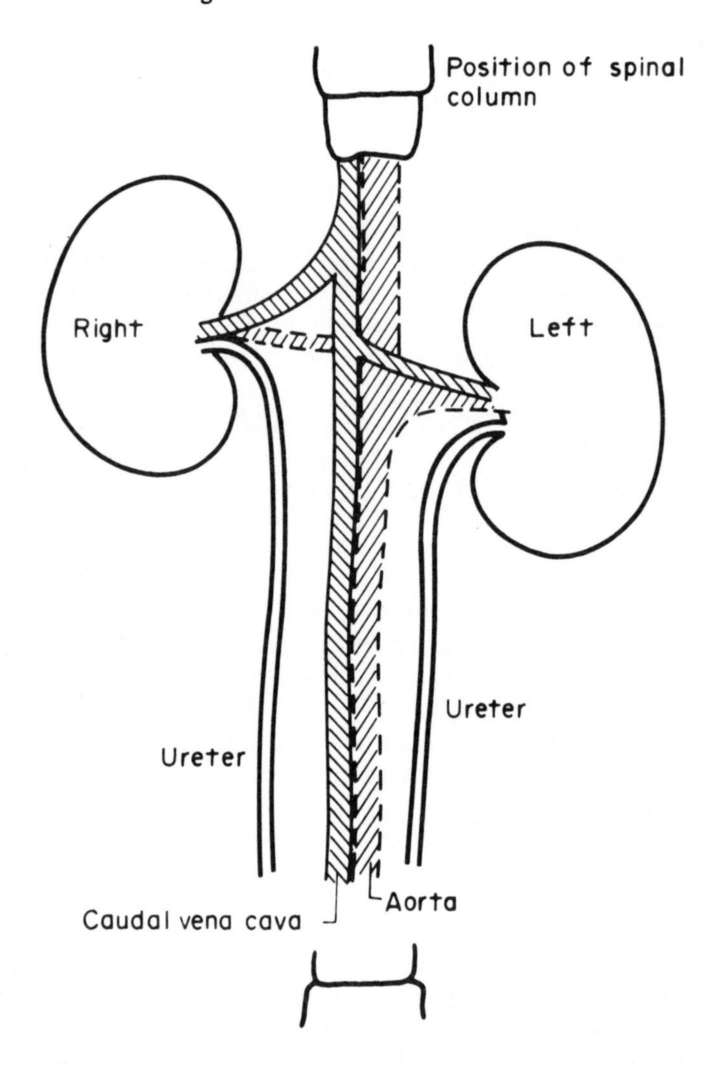

FIG. 3. Gross structure of kidneys.

As the fluid passes down the tubule, much of it is reabsorbed by the tubular wall and returned to the blood. Reabsorption is selective, substances present in excessive quantities in the blood are not reabsorbed, nor are various waste products. Some substances can be actively secreted into the tubule by the cell walls. The tubules converge deep within the kidney in collecting ducts and the remaining contents leave the kidney via a thin-walled tube known as the 'ureter'. A ureter passes from each kidney to the bladder where urine is stored until it can be conveniently voided.

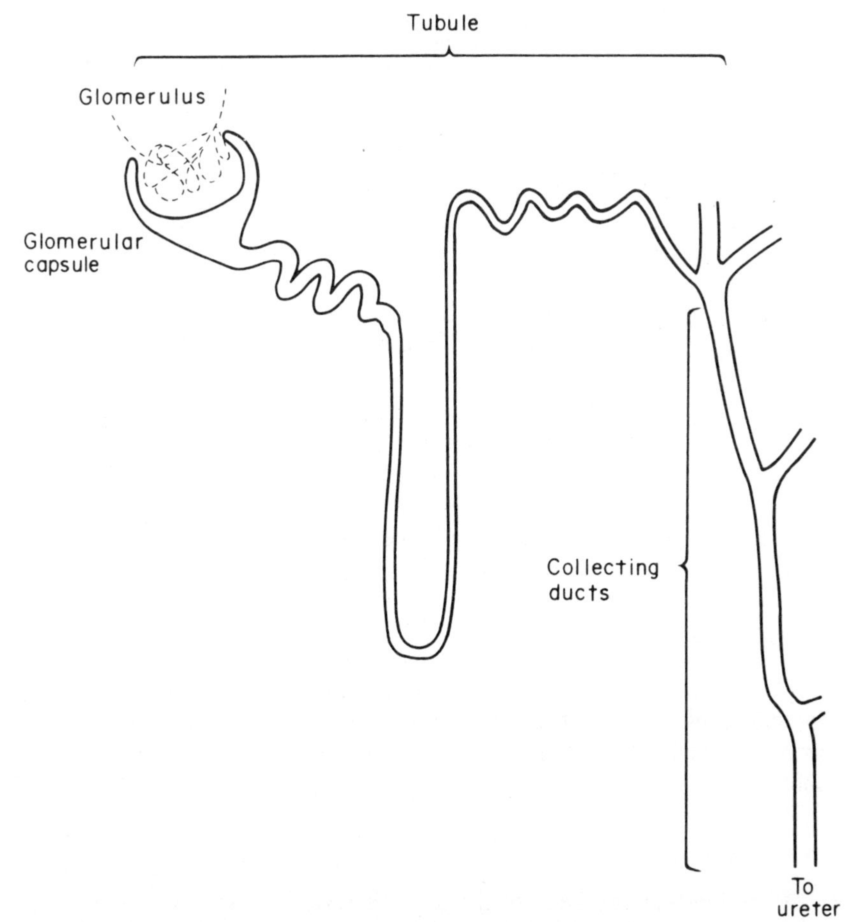

FIG. 4. Structure of kidney tubule (simplified).

Control of the water and electrolytes lost via the kidneys occurs at several levels. There is a rudimentary form of negative feedback; if dehydration results in a loss of ECF volume, blood pressure will drop, therefore forces driving filtration into the tubule will be reduced and less filtrate will reach the tubule. This limits water loss.

Blood pressure may be partly restored by the kidney; when the blood pressure drops, the kidney releases an enzyme called renin which catalyses the conversion of an inactive plasma protein to angiotensin. Angiotensin is a potent hormone which causes constriction of the arteriolar blood vessels, thereby maintaining a minimum pressure despite loss of volume. It also stimulates the adrenal cortex to release aldosterone, another hormone which increases the tubular reabsorption of salt and water. The rate of water absorption is also governed by antidiuretic hormone (ADH) which is pro-

TABLE 4
Metabolic water

Class of food	Water yield on oxidation of 100 g
Protein*	40 g
Fat	107 g
Carbohydrate	55 g

*Not always completely oxidized.

duced by the pituitary gland in the brain in response to elevated concentrations of some of the constituents of blood. ADH acts upon part of the tubule to increase water reabsorption.

The role of the kidney in regulating electrolyte balance, particularly the level of hydrogen ions, may interfere with its role in water balance. Hydrogen ions are produced by many of the body's chemical reactions, and they cannot be allowed to accumulate, as this would change the pH of the body. As these have to be removed from the body in solution, some loss of water as urine is inevitable even in conditions of severe dehydration.

Water Intake

There are several pathways along which water may enter the body. These include drinking water, the water content of foodstuffs, which may be as much as 90% of the food, and the water released during metabolic utilization of these foods.

Water content of foods

When foods are broken down during digestion, water is released together with the other end-products of digestion such as sugars and amino-acids. The quantity of water depends upon the type of foodstuffs; for example, commercial dry dog and cat foods may contain as little as 6% water (although some do contain more) whereas many wet foods contain up to 82% water. Milk contains approximately 88% water, and fresh fish and meat 55–75%. Therefore the amounts of water available to animals from their food can vary 10-fold.

Metabolic water

This is the water produced on chemical breakdown of the nutrients by oxidation in the tissues.

Hydrogen in the food combines with oxygen to produce water. The quantity of water released depends entirely on the class of foodstuff and the degree of oxidation (see Table 4).

TABLE 5
Structure of nutrients

Class	Common forms in food	After digestion
Carbohydrates	polysaccharides (e.g. starch) disaccharides (e.g. sucrose) monosaccharides (e.g. glucose)	monosaccharides ('simple sugars')
Proteins	protein	peptides amino-acids
Fats	neutral fat	glycerol fatty acids some glycerides

Water drunk

Water taken in by drinking is under voluntary control. There are several different feedback mechanisms which stimulate drinking. Receptors in the mouth and throat send signals to the 'thirst centre' in the brain when they are dry. Similarly, certain 'osmoreceptors' feed back to the thirst centre when dehydration causes an increase in the osmotic pressure of the ECF. Severe dehydration resulting in loss of ECF volume and the consequent increase in circulating angiotensin also stimulates the thirst centre.

Fluid requirements obviously vary according to environmental conditions, the animal's physiological state and the water content of any food eaten. The dog adapts its water intake very well according to the water content of its food; the cat less quickly and completely. From a practical viewpoint, *ad libitum* access to fresh drinking water for both cats and dogs will give them the best opportunity to meet their water requirement, particularly in warm conditions and when dry foods are fed.

DIGESTION AND ABSORPTION IN THE DOG AND CAT

The regulation of food intake has already been discussed. However, before the animal's nutritional needs can be met, there is an intermediate step to be considered. This is the breakdown of the large complex compounds in the food into a simple form which can be absorbed from the digestive tract, circulated to the tissues and used by them for maintenance, repair, growth or the provision of energy. This is the role of the digestive system.

There are three major classes of foodstuff which require digestion: carbohydrates, fats and proteins (Table 5). The purpose of digestion is to remove the linkages in the large compounds to free the small units (see Chapter 2). This is achieved by 'hydrolysis', which is the splitting of compounds by water, and it is accelerated by digestive enzymes. Enzymes are organic catalysts, produced by the body, which regulate the progress of most of the biochemical reactions in the body. Digestive enzymes have specific roles, each concerned with a particular step in the breakdown of a particular compound.

The digestive tract of the dog and cat can be thought of as a simple hollow

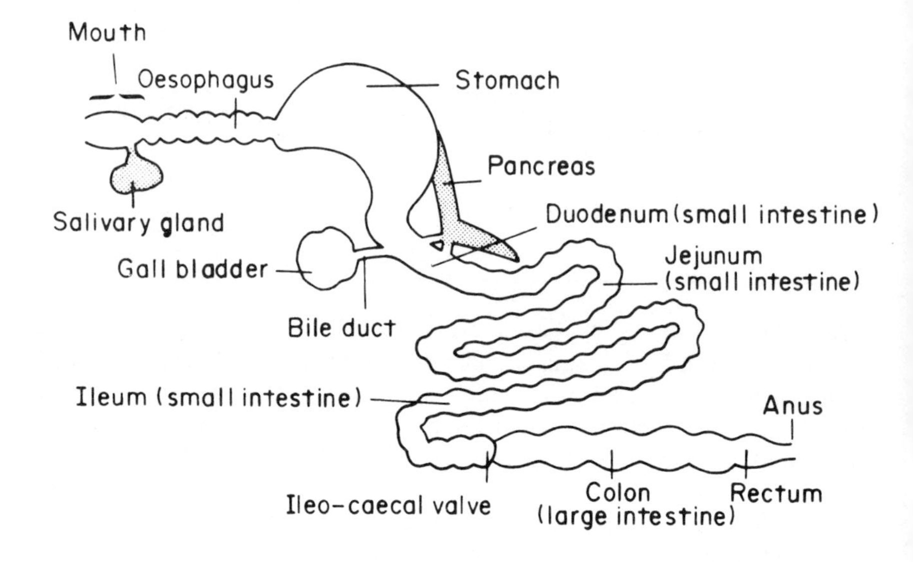

FIG. 5. Simplified monogastric digestive system.

tube, parts of which are differentiated by structure and function. Food passes down from the mouth towards the rectum, reflux being prevented by values between each compartment. Movement of the food is assisted by muscular contractions of the gut wall, often co-ordinated in a sequence called 'peristalsis'. This is where the wave of contraction moves down the gut, taking a bolus of food with it. Figure 5 shows a very generalized monogastric digestive system, which is applicable to both the dog and cat. The different compartments will be considered separately in the order in which they occur anatomically.

Mouth

Once the food has been caught and killed (or at least put down in a bowl within reach of the cat or dog!) the sight and smell of the food evoke the production of saliva by the salivary glands (Fig. 6). This is known as a gustatory response. Pavlov observed that saliva secretion in dogs might even be prompted by a stimulus usually associated with feeding, such as the ringing of a bell at meal times. This saliva production is reinforced when the food is taken into the mouth and taste is added to the other sensations. Saliva is a

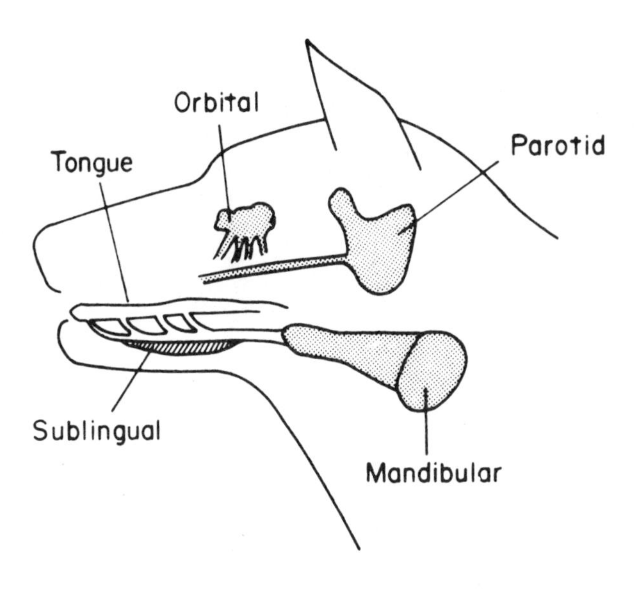

FIG. 6. The salivary glands of the dog.

slightly acid secretion and it contains mucus which is a very effective lubricant and makes swallowing (particularly of dry foods) easier. In some animals the starch digesting enzyme salivary amylase (ptyalin) is present in saliva, but its contribution to digestion is usually negligible.

In cats, chewing allows some mechanical breakdown of the food, but many dogs bolt their food without any chewing. However, if the food is tough, both cats and dogs have the dentition associated with a carnivorous way of life, and are well equipped to tear, gnaw and chew their food.

Oesophagus

Swallowing transfers food from the mouth to the oesophagus, a relatively short tube which leads to the stomach. No enzymes are secreted here, but the cells of the oesophagus will add more mucus to ease movement. The presence of food stimulates peristalsis which pushes the food towards the stomach. At the base of the oesophagus, where it enters the stomach, there is a ring of specialized muscle cells, known as the cardiac sphincter.

Usually in a contracted state, this sphincter is stimulated to relax by the approach of the peristaltic wave, allowing food to pass into the stomach. However, pressure from the stomach side does not cause relaxation, so reflux is unlikely, except in the abnormal circumstances associated with vomiting.

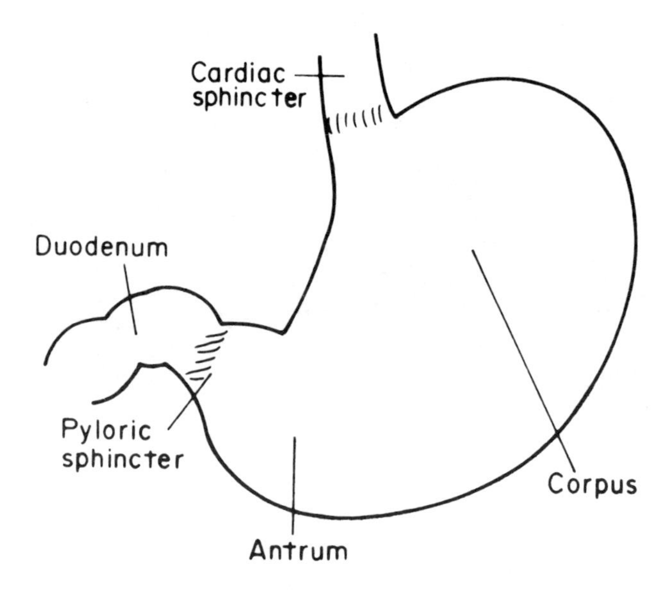

FIG. 7. Stomach of monogastric animal.

Stomach

The stomach has many functions. It acts as a reservoir to allow food to be taken in as meals rather than continuously, it initiates the digestion of protein and it regulates the flow of material into the small intestine. Functionally, the stomach can be divided into two parts, the corpus and the antrum (Fig. 7).

The corpus has very elastic walls which can accommodate large quantities of food without any increase in pressure. The mucosa (epithelium and underlying tissue) of the corpus secretes mucus, hydrochloric acid and proteases. Proteases are protein digesting enzymes, and in the stomach they split the very long protein chains into smaller polypeptides. The major enzyme, pepsin, is secreted in an inactive form, pepsinogen, to ensure that it does not digest the cells in which it is produced. Pepsinogen is converted to pepsin in the presence of hydrochloric acid, which also provides the appropriate environment for the enzymes to function at an optimum rate. The stomach is protected from pepsin by a stream of mucus which lines the walls.

The secretion of acid, mucus and enzymes depends upon the quantity and composition of the food in the stomach and is under both hormonal and nervous control (Fig. 8).

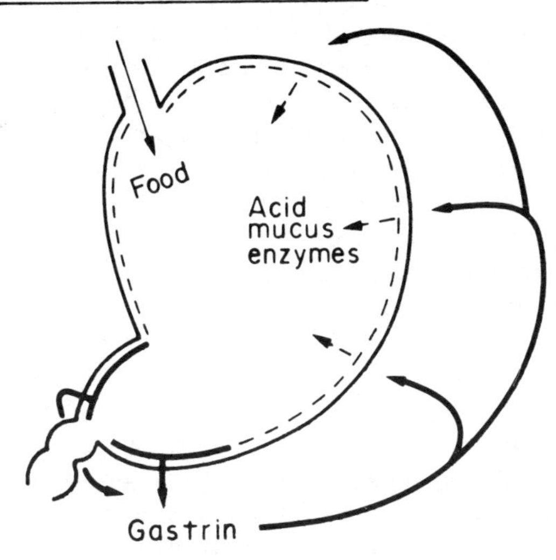

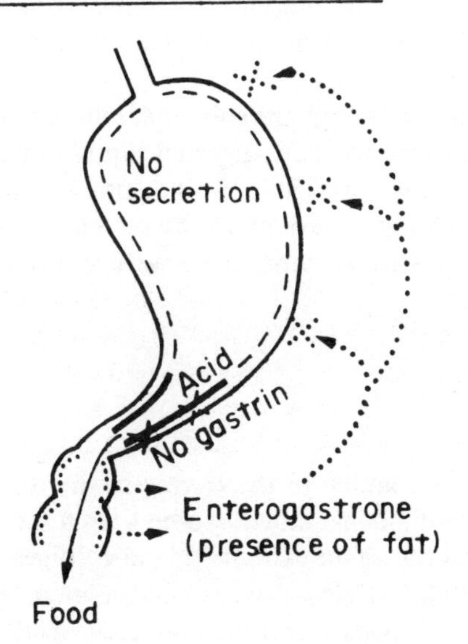

FIG. 8. Control of gastric secretion.

The hormone gastrin stimulates the stomach to produce acid, enzymes and also increases the motility of the stomach. It is produced in the cells of the antral mucosa and released into the blood when the stomach is distended or there is food present. Gastrin travels in the bloodstream until it returns to the stomach where it has its effect on the mucosa of the corpus. The release of gastrin is eventually self-regulating; as the acid secretion causes the pH to drop, the release of gastrin is inhibited. As the stomach empties into the small intestine, the presence of fat stimulates the release of the duodenal hormone enterogastrone, which causes the stomach to cease acid production.

The nervous control of gastric secretion is rather more direct. There is a simple stretch reflex which stimulates secretion, and also a gustatory response to the sight, smell and taste of food, which evokes a secretion rich in proteases and acid, in readiness for the food when it reaches the stomach.

The antral mucosa, by contrast, produces a solution which is alkaline and low in enzymes. This is mixed with the food before entry into the small intestine. Mixing waves originate in the corpus and gradually increase in strength as they reach the muscular antrum where thorough mixing occurs. By this stage the stomach contents form a thick milky liquid known as chyme.

The rate at which the stomach releases chyme into the duodenum (upper part of the small intestine) is influenced by several factors, allowing the optimum conditions for digestion. The mechanisms involved are very simple. At the distal (far) end of the stomach is a tight ring of muscle called the pyloric sphincter, which like the cardiac sphincter is normally constricted. When very strong peristaltic waves approach, the sphincter relaxes and allows chyme to enter the duodenum. The presence of acids, irritants, fat or chyme in the duodenum inhibits peristaltic movement by the stomach and therefore reduces the rate of emptying.

Fluid chyme passes through the sphincter more easily, which means that the passage of well-mixed, partially digested liquid chyme is favoured, particularly when there is none present in the intestine already. This ensures that the small intestine does not receive more chyme than it can cope with efficiently, and also that the gastric enzymes have sufficient opportunity to work in an acid environment.

Small Intestine

Digestion

More enzymes are added to the chyme in the duodenum. Some of these originate from the duodenal mucosa, others from the pancreas. The pancreas is important not only as an exocrine gland in digestion (i.e. a gland which secretes externally), but also as an endocrine gland (a gland which secretes hormones into the bloodstream) in producing insulin. It also secretes large volumes of bicarbonate salts in the gut, neutralizing the acid chyme from the stomach, and providing the optimum pH for the pancreatic and intestinal enzymes. The pancreatic enzymes include inactive proteases, lipases (fat digesting) and amylase (carbohydrate digesting). Intestinal enzymes generally catalyse the later stages of digestion.

Regulation of the pancreatic output is largely under the control of two hormones, secretin and pancreozymin (Fig. 9). They are both produced by cells of the intestinal mucosa and, under certain conditions, released into the bloodstream. Secretin is released in response to acid in the gut and stimulates the release of larger volumes of bicarbonate from the pancreas. In contrast, pancreozymin release is provoked by the presence of partially digested food and it stimulates the release of juices rich in enzymes. The complementary role of these hormones ensures the least wasteful use of the pancreas.

Bile is also added to chyme in the duodenum. Bile is a fluid which is produced continuously by the liver; in some animals (e.g. horses, rats) it trickles directly into the duodenum via the bile duct. However, in other species (e.g. man, dogs, cats) it is stored in the gall bladder, to be released into the duodenum when required. Bile contains bile salts and pigments and various waste products of the liver such as hormone and drug metabolites. Bile salts are not enzymes although they have several important roles in digestion and absorption. The most important is the emulsification of fat, bile acting upon the fat rather like a detergent, splitting it into many tiny globules with a large surface area on which the lipases can act. Some of the lipases are also activated by the presence of bile, in a similar way to the activation of proteases by acid in the stomach.

Secretin, the duodenal hormone, increases the bicarbonate content and rate of flow of bile. Another duodenal hormone, cholecystokinin, causes contraction of the gall bladder and release of the stored bile (Fig. 10).

The small intestine is so called because of its narrow bore, for although its diameter is much less than that of the 'large' intestine, it is several times as long. Digestion is completed in the small intestine, all the digestible protein, fat and carbohydrate being reduced to amino-acids, dipeptides, glycerol, fatty acids and monosaccharides. As these are released they are absorbed, as are the minerals, vitamins and water.

Absorption

Absorption is the transfer of digested material from the lumen of the gut into blood or lymphatic vessels. Although some absorption occurs from the stomach and large intestine, by far the greatest proportion takes place across the mucosa of the small intestine. The surface area over which this can take place is greatly enlarged by folds and numerous small finger-like projections called villi (Fig. 11). In some dogs the surface area of the small intestine may be equivalent to the floor of a small room. Independent mixing movements of the gut wall and villi ensure that there is a good supply of materials to the epithelial surfaces, and the dense capillary network ensures that there is no accumulation of absorbed nutrients which might hinder further absorption.

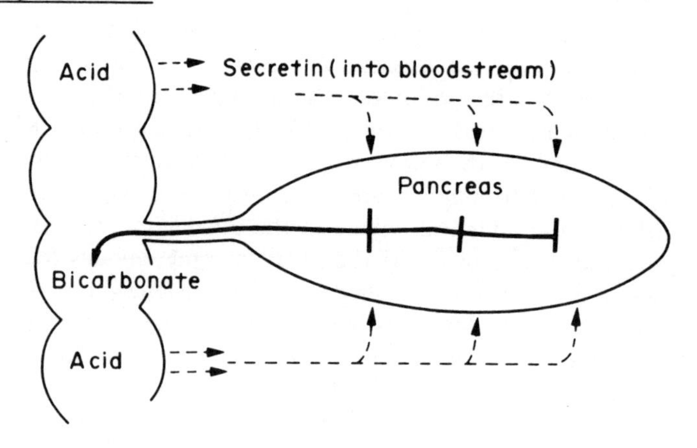

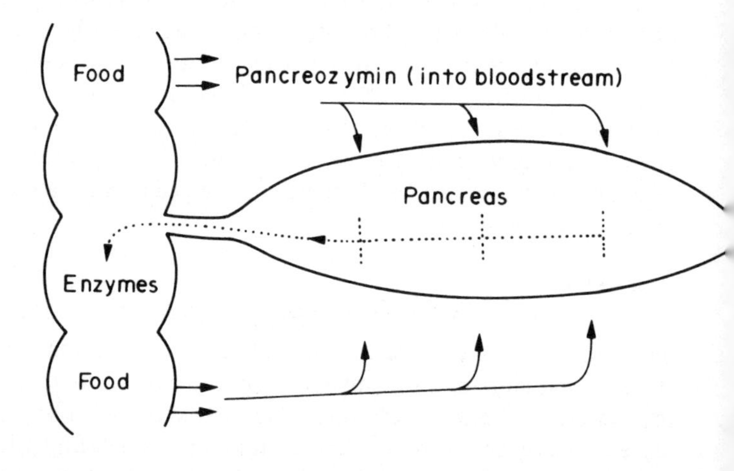

FIG. 9. Regulation of pancreatic secretion.

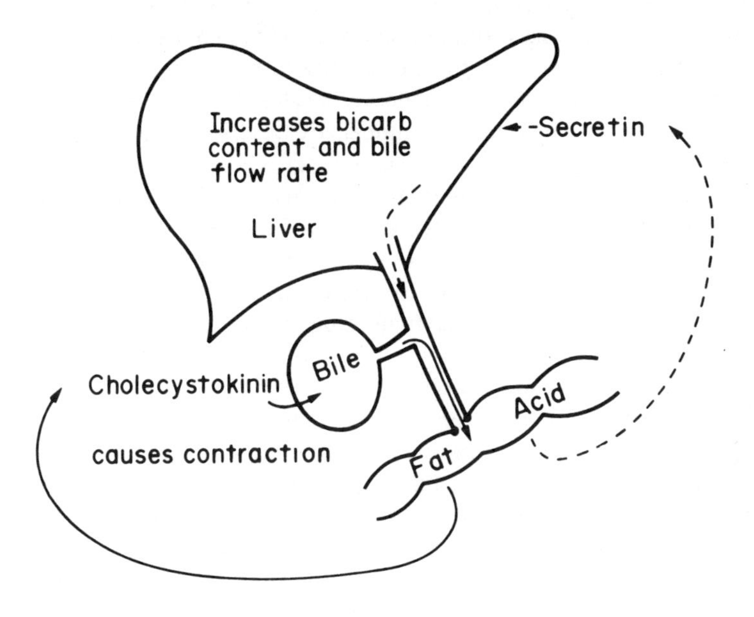

FIG. 10. Bile production and release.

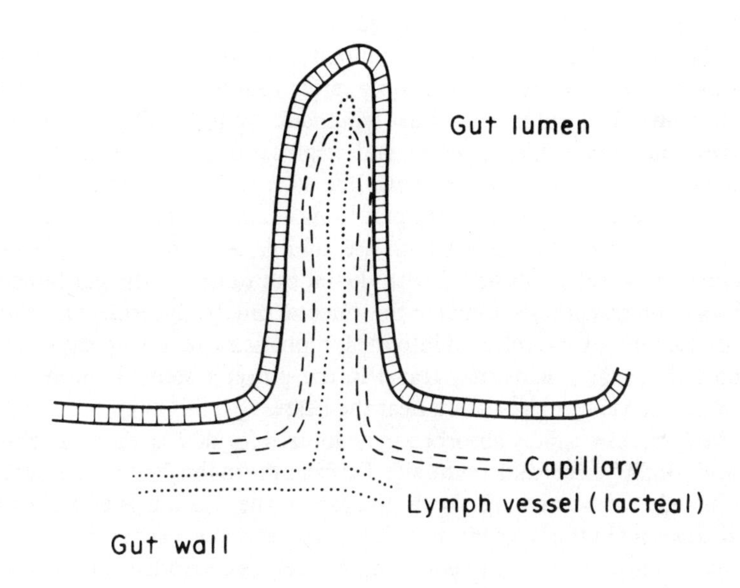

FIG. 11. Intestinal villus (simplified).

There are several different ways in which nutrients are absorbed. Absorption may be 'passive', according to the concentration or osmotic gradient, or 'active' requiring the expenditure of energy to run 'pumps' across cells or membranes. Amino-acids and monosaccharides demonstrate some passive diffusion but this is very limited. Amino-acids are absorbed actively by several different transport systems into the mucosal cells, then diffuse into the bloodstream. Some of the digested protein is absorbed as dipeptide (two amino-acids) by active systems, the dipeptide bond being broken within the villus cell wall, and the two amino-acids released singly into the bloodstream. Newborn animals are also able to absorb intact proteins (e.g. maternal antibodies in colostrum) by an enfolding action of the cell, known as pinocytosis.

The active uptake of monosaccharides is linked to a carrier complex which depends on sodium uptake. Other minerals (e.g. calcium) are also linked to the movements of monosaccharides. Sugars and amino-acids are absorbed into the villus capillaries and from there converge on the portal vein which shunts the blood through the liver before returning to the heart for recirculation. The liver converts much of the absorbed glucose into glycogen (regulated by the level of circulating insulin) and stores it until a drop in blood glucose calls for the conversion of some of the stored glycogen to glucose (regulated by glucagon). The circulating level of glucose has to be maintained in order to provide a ready supply of available energy for the tissues, particularly in the brain. Amino-acids circulate in the blood and are absorbed by the cells as required; surplus amino-acids are either converted to others as needed or broken down in the liver to urea, which is then excreted by the kidney.

The absorption of fat differs from that of protein and carbohydrates; fatty acids and glycerol are absorbed only rarely into the villus capillary, the bulk being absorbed into the villus lymphatic system. The products of fat digestion, fatty acids, glycerol and triglycerides, are insoluble in water. However, they form micelles with bile salts and lysolecithin and these can disperse freely in the fluid in the gut. Long chain fatty acids associate with bile salts to form cholic acids which are soluble in water. The bile salts and lysolecithin are not absorbed with the fat but return to the gut lumen. After absorption resynthesis occurs in the mucosal cell, triglycerides and phospholipids are formed and released into the lymphatic system, whereas glycerol and short chain fatty acids may travel in the portal system. Lymph eventually rejoins the venous circulation near the heart.

Minerals are usually absorbed in an ionized form. The means of absorption varies slightly according to the site; for example in the jejunum sodium uptake is linked to the active uptake of glucose, in the ileum it is entirely an active process, and in the large intestine it is very active (i.e. can operate against very strong concentration gradients) and entirely independent of glucose movement. The absorption of minerals depends on their levels in the body (which will influence concentration gradients) and on various hormonal factors.

Water soluble vitamins (B group) are usually absorbed passively but there may be some active absorption. Vitamin B_{12} can only be absorbed after binding to a protein known as the intrinsic factor, which is produced by the gastric mucosa.

Fat soluble vitamins (A, D, E and K) are made soluble by combination with bile salts and this aids their absorption. Where there is normal fat digestion and absorption, there should be normal fat soluble vitamin uptake.

Water is absorbed passively by diffusion down an osmotic gradient. The bulk of water is absorbed in the small intestine, with a little in the stomach and large intestine. If water uptake is impaired, dehydration may occur very quickly as all the watery secretions in the gut will be lost, in addition to water drunk and that contained in food.

The Large Intestine

The contents of the gut enter the large intestine through the ileo-caecal valve. Little of the food and water taken in at the mouth reaches the large intestine, that which does is mostly destined for evacuation as faeces.

The large intestine has no villi, so its surface area is limited, and although it is capable of taking up water and some electrolytes, it has none of the transport mechanisms needed for organic nutrients. Water is absorbed differently here, being drawn into the intercellular spaces by the setting up of gradients. The degree of absorption is affected by the fluid status of the body, reflected by the presence or absence of the hormones aldosterone and angiotensin. The ileum and colon (part of the large intestine) are particularly sensitive to these. There is also a slight inhibitory action of secretin, gastrin and pancreozymin on water uptake.

The bacterial colonies resident in the large intestine are able to partially digest some of the protein and fibre residue. The products of this digestion give the faeces their characteristic smell and colour. Any residues left undigested, together with water, minerals and dead bacteria, are stored in the rectum until evacuation. Defaecation is usually under voluntary control, involving the relaxation of an anal sphincter, but diarrhoea or illness may override this control.

An understanding of physiology of the digestive tract helps in the interpretation of gastro-intestinal disease. Thus poor absorption of water, either from some impairment of the mechanisms, or by too rapid a transit time, results in diarrhoea. If too much absorption occurs, the faeces become hard, difficult to evacuate and constipation occurs. Vomiting may be caused by toxins or poisons which irritate the stomach wall, or by disease of the pyloric sphincter. Swallowing of foreign bodies may also cause vomiting.

Persistent diarrhoea and vomiting may prove fatal because of the loss of inorganic ions and the effect of dehydration. They may be an indication of serious damage or disease in some part of the digestive tract. However, occasional nausea and loose faeces may be caused by nothing more serious than a sudden change in dietary regime, or a period of overfeeding.

CHAPTER 4

A Balanced Diet

ANNA L. RAINBIRD

Dogs and cats both belong to the order Carnivora, a group of mammals distinguished by the arrangement of their teeth and by a feral, predatory way of living. It is likely that they evolved on a diet consisting mainly of other animals which they caught and killed. This does not mean that meat or other parts of animal bodies are the only suitable food for present day domesticated dogs and cats. Indeed it is quite possible to maintain dogs and cats in good health on diets which do not contain any meat as such, but consist only of nutrients provided as pure chemicals with 'protein' supplied as amino-acids only or as mixtures of amino-acids with isolated protein of vegetable origin. Such purified or semi-purified diets made from sugar and starch, fat, vitamins, minerals and amino-acid mixtures, are not foods as we generally recognize them and are mentioned only to emphasize the fact that cats and dogs can be provided with adequate supplies of nutrients in a variety of ways and that no single food is necessarily the 'best' or most suitable for them.

In considering what it is that makes some foods clearly unsuitable for dogs and cats and others more or less suitable, we have to be aware of several factors. Some of these depend on the physiology of the animal, some on the food itself and some on the expectations which people have about foods.

Animals eat to provide themselves with the energy they require. At the same time they must also obtain all the other essential nutrients they require, in the correct amounts and the correct proportions. Any material which can provide nutrients or energy is a potential food source but the nutrients in food only become available to the animal through the processes of digestion and absorption which take place in the alimentary canal or gut. So unless the digestive system of the dog or cat is capable of breaking down and absorbing the nutrient content of a potential food source, it is not going to be suitable as a food for those species. In general, dogs are able to digest food more efficiently than cats, with the exception of some highly digestible raw materials (Kendall, 1981). However, both species are unable to digest plant cell wall materials, commonly known as dietary fibre or roughage. Dietary fibre consists mainly of cellulose and hemicelluloses. Foods containing high levels of dietary fibre usually have a low digestibility but this can be improved by processing. Because of other properties, food rich in dietary fibre may also have a role to play in the prevention or treatment of certain conditions in dogs, e.g. diarrhoea, constipation.

Cats are true carnivores and must have foods of animal origin in their diet. Cats have a total lack of, or produce inadequate amounts of, some enzymes possessed by dogs and other mammals which render them incapable of using normal metabolic pathways to synthesize certain nutrients within the body. These have to be supplied preformed in the diet. Examples are vitamin A, arachidonic acid (an essential fatty acid) and taurine, an amino-sulphonic acid. These essential nutrients are not found in vegetable materials and thus, at least part of a cat's diet must be food of animal origin. Both cats and dogs show a distinct preference for animal protein and animal fat.

Food cannot just be considered as a source of nutrients and energy. Other characteristics and attributes of food must be considered when examining their suitability as a component of a diet. Dog and cat owners are concerned to maintain the health, activity and life of their pets for as long as possible. Feeding is seen as an occasion to be enjoyed by both, and unless food is eaten with obvious enjoyment, some of the reward and good feelings of being a provider may be lost to owners and their opinion of the foods' suitability lowered. It is inappropriate to evaluate foods only in terms of nutrient content and cost. Other factors such as safety, apparent enjoyment by the dog or cat, acceptance, suitability for feeding in the home, which involves smell and appearance, keeping quality, convenience in purchasing, storage and preparation are all important to varying degrees.

Safety as a quality of a food is usually taken for granted but it is important that foods do not contain toxic components or contaminants including food spoilage organisms. Cooking and processing of food material is done mainly to ensure safety. It may also improve the appearance, taste, texture and digestibility of some foods, but the main objective of cooking is to make food safe to eat. The application of heat kills bacteria and moulds (food spoilage organisms) and also destroys most of the toxins or poisons which they produce. It kills parasites such as the eggs, larvae and encysted larvae of worms, and destroys many of the naturally occurring toxic materials present in some vegetable foods, for example, goitre-causing substances in some brassicas, trypsin inhibitors in soya beans and the cyanogenetic glucosides in tapioca or cassava. Cooking also destroys bacteria like *Salmonella* and *Botulinum* which can cause severe food poisoning. It also improves digestibility of foods for dogs and cats, particularly of starchy, vegetable foods like cereals by bursting the starch granules and exposing them to the action of digestive enzymes.

Cooking usually softens and tenderizes meat containing large amounts of collagen but probably does not increase the digestibility of proteins to any marked degree. Overcooking can be harmful in destroying protein structure, some vitamins, and in causing loss of vitamins and minerals by leaching into cooking gravies or liquors.

Palatability and acceptance by dogs and cats are attributes which are hard to describe but easy to recognize. Their importance is paramount as unpalatable, unacceptable food which remains uneaten obviously has no nutritional value. Food which animals appear to eat with less than avid enjoyment and gusto is likely to be rejected by the owner, unless factors such as cost and convenience are more important.

In the domestic situation the needs of the owner are important — does the food need to be cooked before feeding? Is it pleasant and convenient to handle and store before and after cooking or messy and unpleasant like green tripe? Owners have prejudices about foods and so some foods which are quite acceptable and useful for one person, can be completely unacceptable to another.

The total diet of a dog or cat may consist of a single food fed every day or it may be any number of combinations of mixtures of foods fed regularly or intermittently. It will be satisfactory if it is sufficiently palatable to be eaten in quantities which supply enough available nutrients to meet the needs of the animal. Foods which are good sources of some nutrients but deficient in others, will be completely unsuitable as the sole diet for a cat or dog but may be ideal for feeding part of a mixed diet.

What then are suitable foods for cats and dogs? This is a very difficult question and depends upon the meaning applied to the word 'suitable'. Suitability relates to fitness for the purpose intended. In the context of food it covers a very wide range of qualities, for example food which may be quite suitable for an adult dog may be quite unsuitable for a young kitten. As a rule of thumb it is almost true to say that if a food is suitable for people, it is also suitable in some degree for cats and dogs.

This does not help in choosing foods for cats and dogs and so the following sections of this chapter attempt to describe the characteristics of several classes or types of foods in terms of nutrient content and availability, palatability and other aspects which allow a judgement to be made on their place in the construction of satisfactory feeding regimens.

MEAT AND MEAT BY-PRODUCTS

Meat, as we usually think of it, consists of the *muscle* tissue of animals together with the associated intramuscular fat, the connective tissue of muscle sheaths and tendons, and blood vessels. It may also include various amounts of subcutaneous fat overlying the muscle, and marbling fat which is contained within the lean meat but between muscles. The relative proportions of muscle fibres and connective tissue have a strong influence on the toughness or texture of the meat. The real differences in nutrient content between muscle meat from different parts of the carcase depend on the proportion of fat present. Lean muscle devoid of fat has much the same proportions of water and protein, 75% water, 25% protein, no matter whether it is from different parts of the same animal or even from different animals such as cattle, sheep, pigs or poultry. A most comprehensive collection of data on the nutrient content of meats and other foods is contained in Paul and Southgates' 1978 revision of McCance and Widdowson's 'The Composition of Foods'.

Raw lean meat (including intramuscular fat) from pork, beef, lamb, veal, chicken, turkey, duck or rabbit, is very similar in composition and has average water, protein and fat contents of about 70–76%, 20–22% and 2–9%, respectively. Fat is the major variable, the 'white' meats (poultry, veal and rabbit) having rather less fat (2–5%) than lamb and pork (7–9%), but being very little different from lean beef. Because the proportions of lean to fat vary widely between different joints of the same carcase and between different animals, it is impossible to make accurate predictions of the protein, fat and energy content of an individual piece of meat using food composition tables. Lamb and pork are generally fattier than beef carcases. Boneless edible meat from a modern beef animal contains on average about 24% fat with individual joints having from as little as 5% (lean meat) up to 26% (fore-rib and brisket). Breast of lamb or loin chops may contain up to 36% fat and the whole edible carcase averages about 30–32%. The leg may contain as little as 18% fat. The protein content of beef ranges from 16–18% for rib, up to about 20% for steak.

The protein quality of meat from all the animals and birds mentioned is of high value. There is therefore little to choose in terms of nutrient supply between meat from different meat animals, once allowance has been made for the fat content and this varies as much or more between parts of the same animal as it does between species. Fat from poultry and pigs is likely to be more unsaturated (that is, contains more double bonds) than that from beef and sheep but, in practice, this has little effect on the digestibility or availability to dogs and cats.

Offal meats or meat by-products such as liver, kidney, tripe, melts, lights are generally similar in nutrient content, no matter what species they are from. However, there are large differences in the nutrient content of different offals, e.g. liver has a very different nutrient content compared with tripe. They tend to have a variable fat and vitamin content depending on the food fed to the animal. Table 6 shows some typical water, protein and fat contents of various meats and offals with calcium, phosphorus and energy values. Tables of food composition should be consulted for more detailed information. For all practical purposes, meat is devoid of carbohydrate because energy reserves are present mostly as fat. Muscles do contain small amounts of glycogen but this is rapidly depleted after slaughter and so the content of carbohydrate is negligible.

All meats, both muscle meat and offals, are very low in calcium and have very adverse calcium:phosphorus ratios of 1:15 to 1:26. This leads to very

TABLE 6

Typical nutrient content of some meats and meat by-products

	Water g/100g	Protein g/100g	Fat g/100g	Calcium g/100g	Phosphorus g/100g	Energy kcal/100g
Raw lean meats						
Pork	71.5	20.6	7.1	0.008	0.20	147
Beef	74.0	20.3	4.6	0.007	0.18	123
Lamb	70.1	20.8	8.8	0.007	0.19	162
Veal	74.9	21.1	2.7	0.008	0.26	109
Chicken	74.4	20.6	4.3	0.01	0.20	121
Duck	75.0	19.7	4.8	0.012	0.20	122
Turkey	75.5	21.9	2.2	0.008	0.19	107
Rabbit	74.6	21.9	4.0	0.022	0.22	124
Offals						
Udders	72.4	11.0	15.3	0.26	0.24	182
Fatty lungs	73.1	17.2	5.0	0.01	0.19	114
Sheep lungs	76.0	16.9	3.2	0.01	0.20	96
Brains	79.4	10.3	7.6	0.01	0.34	110
Stomachs (pig)	79.1	11.6	8.7	0.03	0.11	125
Spleen	75.9	17.0	6.5	0.03	0.22	126
Kidney (beef)	79.8	15.7	2.6	0.02	0.25	86
Heart	70.1	14.3	15.5	0.02	0.18	197
Heart (trimmed)	76.3	18.9	3.6	0.005	0.23	108
Liver (fresh)	68.6	21.1	7.8	0.001	0.36	163
Green tripe	76.2	12.3	11.6	0.01	0.10	154
Dressed tripe	88.0	9.0	3.0	0.08	0.04	63

severe problems of bone under-mineralization if meats are fed as the major part of the diet without proper supplementation. All muscle meats and most offals and meat by-products are deficient in vitamins A and D. Liver and, to a lesser degree, kidney are good sources of these vitamins. Liver can in fact contain so much vitamin A (retinol) that cats in particular have had very serious ill effects when fed almost nothing else. Meats are generally good sources of high-quality protein, fats, iron and some of the B group vitamins, particularly niacin, thiamin, riboflavin and B_{12}. They are highly palatable to dogs and cats and generally of high digestibility, which means that their nutrient content is readily available. When properly supplemented with sources of calcium, phosphorus, iodine and vitamins A and D they make excellent foods.

Among the meat by-products are materials like blood, bone, whole rabbit or chicken carcases, pig's and sheep's heads and feet, poultry, carcases from which most of the flesh has been used in the preparation of human foods. Because they may include bones, they often have high calcium and phosphrous contents, which helps complement the deficiencies of meats like lungs and livers. They are less suitable for use in home-made foods because they are difficult to handle and prepare but are used in the manufacture of prepared foods.

FISH

Fish are commonly divided into fatty fish and white fish. White fish like cod, haddock, plaice, whiting and sole usually contain less than 2% fat, whereas the oily or fatty fish like herring, mackerel, pilchards, sardines, sprats, tuna, salmon, trout and eels may have very much more, between 5 and 18%, according to season of the year or stage of maturity of the fish when caught. In general, white fish are very similar to lean meat in composition. The protein is of similar high quality and the vitamins A and D are generally absent or present only in trace amounts. But fish muscle does contain adequate amounts of iodine and because bones are frequently consumed with the flesh of fish, the calcium and phosphorus content is much better balanced. Filletted fish with bones removed is seriously deficient in calcium and phosphorus. The flesh of oily fish does contain vitamins A and D and the livers of fish like cod and halibut are particularly rich sources of these fat soluble vitamins. Whole fish, including the bones (if made safe by cooking or grinding up), are better balanced sources of nutrients for dogs and cats than most meats.

Fish are usually less palatable than meats but nevertheless are mostly quite well accepted by these animals, but their smell and appearance may be less acceptable to some dog and cat owners. Like meat, fish can contain parasites and should be cooked before being used as food. In addition, some fish muscle contains an enzyme, thiaminase, which breaks down the vitamin thiamin. This enzyme is destroyed or inactivated by heat and provides another reason for cooking fish prior to feeding it.

DAIRY PRODUCE

Cream, skimmed milk, whey, yoghurt, cheese and butter are all milk products which usually contain, in a more concentrated form, some of the nutrients of the original milk. They are usually more palatable to dogs than cats but may be well liked by both species. A small number of cats and dogs may be unable to tolerate more than a minimum intake of milk sugar, lactose, which can result in diarrhoea in those animals which have insufficient lactase, the digestive enzyme which breaks down lactose into its component parts. These animals are easily identified and although they may be able to eat cheese and butter, should not be given milk or other dairy products.

Milk contains most of the nutrients needed by cats and dogs but is a poor source of iron and vitamin D. The riboflavin content is sensitive to sunlight and most of it will be destroyed together with the vitamin C (not a dietary essential for cats and dogs) if it is exposed to sunlight for more than an hour or two. It is a good source of readily available energy, protein of high quality, fat, carbohydrate, calcium, phosphorus and several trace elements, vitamin A and B complex vitamins. Whole pasteurized milk contains about 65 kcal/100g with 3.3 g protein, 3.8 g fat, 4.7 g lactose, 0.12 g calcium and 0.095 g phosphorus.

Skimmed milk is milk from which most of the fat and fat soluble vitamins have been removed as cream. It therefore has almost no fat or vitamins A, D and E and a higher concentration of protein and lactose. Dried whole milk or dried skimmed milk are more concentrated forms of their liquid counterparts. Milk from Channel Island breeds of cow contains rather more fat and protein than the average and has an energy content of about 75 kcal/100 g. Otherwise it has similar properties. Goats' milk is very similar to cows' milk in composition and is of no extra value to cats and dogs.

Yoghurt is made by fermentation of whole or skimmed milk with lactic acid producing bacteria. It may have sugar added to it or some fruit pieces, or extra dried skimmed milk. Usually it contains the same nutrients as the milk or skimmed milk from which it is made but has more energy if sugar has been added. Cream is rich in fat and fat soluble vitamins. Cheese is made by the coagulation of milk protein using rennet in acid conditions. Most of the protein, fat, calcium and vitamin A content are retained in the cheese while the milk sugar and B vitamins are removed in the whey. Most cheeses have similar amounts of protein and fat except for those like cottage cheese which are made from skimmed milk and so contain almost no fat. Cream cheeses contain most fat.

All dairy produce usually has a high digestibility and the nutrient content is readily available to the animal. Apart from those individual animals with a poor tolerance for lactose, these foods are excellent sources of the major essential nutrients and many minor ones.

EGGS

Eggs are good sources of iron, protein, riboflavin, folic acid, vitamin B_{12}, and vitamins A and D. They also contain appreciable amounts of most other nutrients except vitamin C and carbohydrates. Contrary to a popular belief, differences in nutritional value brought about by the system of egg production, i.e. battery, deep litter or free-range are very small. They may affect the folic acid and vitamin B_{12} content but very little else and the differences are probably not important in relation to total diet. Deep-coloured yolks and brown shells do not indicate rich supplies of vitamin A or its precursors because the orange pigment mainly responsible for yolk colour is not carotene.

Eggs are usually eaten without the shell and yet the shell provides a very good source of calcium being made mostly of calcium carbonate and protein. However, most dogs and cats are reluctant to eat egg shells and it may be difficult to get them to eat eggs with shell, unless the shells have first been ground to a powder. Even without the shell they are excellent food. The white is almost all protein and water with trace minerals and some B vitamins. Most of the B vitamins and all of the fat soluble ones are found in the yolk which contains more fat and protein and much less water than the white. Eggs are a poor source of niacin.

Raw egg white contains avidin, a protein-rich material which makes biotin (a vitamin) unavailable to dogs and cats. It is therefore not good practice to feed raw egg whites as a frequent part of the diet. Cooking destroys the biotin-binding effect. The dangers of feeding raw egg whites are often overstated and it is unlikely that the feeding of one raw egg a day to the average dog will have any adverse effects. Cooked egg white is better digested and there is no advantage from feeding raw eggs, particularly as they are often associated with diarrhoea in some dogs.

CEREALS AND CEREAL BY-PRODUCTS

Cereals are the seeds of grasses. Generally they include grains like wheat, barley, oats, rice, rye, maize or corn and some sorghums. Cereal grains consist of the germ or embryo surrounded by a starchy endosperm whose function is to provide storage carbohydrate (starch) and some protein (gluten) to support the growth of the germ. This endosperm is itself surrounded by an aleurone layer, a thin layer of cells rich in protein and phosphorus outside which is the tough outer seed coat. Milling of cereals separates the various layers so that bran contains the tough outer coat rich in polysaccharides, cellulose and hemicelluloses (dietary fibre), flour which is largely endosperm, and germ which is the embryo.

The whole grains of the common cereals wheat, oats, barley, rice, maize, contain about 12% moisture, from 9 to 14% protein, 2–5% fat and about 70–80% carbohydrate as starch. Wheat and oats and barley have a higher protein content and less fat than maize and rice. Generally cereals are used as a source of energy for dogs and cats but they also provide a significant proportion of the protein to the total diet.

There is little to choose between the cereals in the quality of the protein they provide. They do contain substantial amounts of other nutrients, particularly thiamin and niacin. The bran and offal portions obtained by separating the tough seed coat are good sources of dietary fibre and of phosphorus, but unless cooked before feeding, much of the phosphorus will be unavailable because it is present in a complex called phytate. Cooking makes this more available. The high fibre content makes bran useful as a bulking agent when diets of low nutrient content are required but depending on the extraction rate of the milling process, bran can contain up to 200 kcal/100 g. Bran is also useful in that its inclusion in the diet tends to have a good effect on faeces consistency, reducing the likelihood of constipation or diarrhoea. Wheat germ or other cereal germs are rich in thiamin, protein, have more fat and vitamin E.

In the average household, cereal products are much more available than straight whole grains or meals. In the United Kingdom, flours made from wheat are required to be fortified with iron, thiamin and niacin, except for wholemeal flour which consists of the ground whole grain. They must also contain extra calcium and so foods like bread and biscuits baked mainly from flour, are good sources of these nutrients. Bread has about 8 or 9% protein, 2% fat and 45% carbohydrate. It provides many minerals and some B vitamins.

Rice is available as white, long grain rice or as short grain or 'pudding' rice. These are the whole grains minus their outer coats and so consist mainly of starch (about 85%), with low levels of vitamins, minerals, fat and protein (7%). Rice is not readily eaten by dogs or cats unless cooked and then is usually mixed with more appetizing foods before being fed. Because the gluten in rice is different from that in wheat, it is often used as a major energy source when investigating food allergies and in the dietary management of dogs with wheat gluten enteropathy.

Breakfast cereals are usually made using a variety of processes from the whole grains of oats, wheat and maize, although rice and bran are also popular. The processing usually involves fairly extensive heat treatment in flaking or popping and this destroys most of their thiamin. Many breakfast cereal products now have additional thiamin and other B group vitamins added after processing. Digestibility is probably improved because of the cooking they receive. Sago or tapioca are considered as cereals by many people, although in fact they are made from the starchy root of the cassava plant. This has much less protein than rice and is almost entirely starch with small amounts of minerals and only traces of vitamins. For practical purposes it should be treated as contributing only energy. Cereals are not particularly palatable to cats and dogs, even when moistened with water and usually need to be fed as part of the diet only. Their nutrient content is of lower digestibility than many other foods particularly if the cereals are not finely ground or cooked. Fine grinding or cooking markedly increases the digestible energy and digestible dry matter values. This is mainly because of the effect in improving the digestion of starch.

FATS AND OILS

Fats and oils include obvious materials like butter, margarine, lard and dripping, the visible fat in meats and the invisible fats in substances like nuts, lean meats and other foods. Oils are distinguished from fats only by their melting point; they are liquid at room temperature, fats are solid. Fats and oils are made up of triglycerides, consisting of three fatty acids joined to a single glycerol molecule. The differences between the various fats are basically because of the different fatty acids they contain.

Fats may be classified as containing largely saturated fatty acids or unsaturated ones. Most fats contain both kinds but the proportions differ. The unsaturated fatty acids — linoleic, linolenic and arachidonic acids (with 2, 3 and 4 double bonds respectively) — are known as essential fatty acids (EFA) because they are required in small quantities for optimal health and cannot be synthesized by the cat or dog from other fats. Dogs like man, can convert linoleic acid into linolenic and arachidonic acids by desaturation (addition of double bonds) and so need only to be fed sources of linoleic acid. Cats on the other hand appear to have some enzyme deficiencies which make the conversion from linoleic very slow or not possible. Linoleic acid is widely distributed in vegetable seed oils and occurs in small amounts in some animal fats, particularly pork fat and chicken fat. Arachidonic acid occurs only in small amounts in some animal tissue fats. It is not usually present in subcutaneous and storage fats and is easily destroyed by heat, so lard, dripping and rendered fats generally contain almost none.

All fats yield a similar amount of energy, about 2.25 times that obtained from protein or carbohydrate and their nutritional value in other respects depends largely on their origin and vitamin content. Vegetable fats are seed fats. Oil seeds contain 20–40% fat. This is almost pure fat with only traces of minerals and no vitamins apart from vitamin E. Wheatgerm oil is a good source of E activity followed by sunflower and cottonseed oils.

Vegetable fats are usually better sources of unsaturated fats like linoleic acid than animal fats; sunflower oil, soya bean oil and corn or maize oil have about 50% of this fatty acid, safflowerseed oil contains even more — 65–70% — but coconut oil contains hardly any and olive oil only about 10%.

Milk fat and beef tallow contain similar low levels of 2–4%. Fish oils are a good source with over 20% of linoleic acid. They usually contain many other unsaturated fats as well. Most animal fats contain only trace amounts of B vitamins but cod liver oil, butter and margarine are good sources of vitamin A. Margarine and cod liver oil, or other fish oils like halibut liver oil, are also good sources of vitamin D. Margarine has no inherent vitamins from its component vegetable fats but in the United Kingdom must be fortified with vitamins A and D. The very high levels of vitamins A and D in fish liver oil make this suitable for only limited inclusion in the total diet.

Fats, particularly animal fats, are greatly liked by cats and dogs. They add flavour and palatability to other foods. They also bind together dusty foods like cereal meals and are more than just a source of energy and vitamins A and D. Fats are normally highly digestible and slow down the rate of stomach emptying. Fats also help give animals a feeling of satiety after meals. Cooking fat which has been used several times for deep fat frying should never be given as food. It is likely to contain peroxides and other toxic materials which can be harmful if fed to cats and dogs.

OTHER ANIMAL BY-PRODUCTS

Many kinds of by-products from the slaughtering industry are available as foods for cats and dogs. Many are obtained fresh or frozen direct from the abattoir without processing and most of these have been mentioned previously. Others are first processed or are the result of processing of parts of the animal. Among these are dry meals such as blood meal, meat-meal, meat and bone meal, greaves (the dry remnants remaining after fat rendering).

These products form part of the animal feeding stuffs trade and are usually produced to a specified level of protein, fat and ash content. The processes by which they are produced have variable control of the heat treatment given. Protein quality and availability can be very variable as can the ash or mineral content. They provide large amounts of protein from as little as 40% in some meat and bone meals up to 75 or 80% in a high protein meat meal. They can be very good value for money but require careful monitoring. Digestibility by dogs and cats is variable and protein digestibility may be as high as 90% or as low as 70%. They are often quite palatable and improve acceptance when mixed with cereals. Their chief use is as part of dry manufactured foods. The only product of this which finds much application in home-made foods is sterilized bone flour. It must be sterilized to reduce any risk of disease, particularly salmonellosis and foot and mouth disease. Bone flour contains about 32% calcium and 14% phosphorus and is a very good calcium-phosphorus supplement for meat; 15 g will adequately supplement 1 kg meat or meat by-product.

Fish meals are made by drying the whole bodies and offal of white fish or from the whole bodies of oily fish such as herring, mackerel or sprats when they are in over supply. They are usually very good sources of high quality protein for use in foods for cats and dogs. The protein and fat content depend upon the fish used. White fish meal usually contains 66–70% protein with fat levels of about 6–8% and meal from oily fish may have up to 80% protein and 8–10% fat. Both kinds are rich in minerals as one would expect from the composition of whole fish. They are usually palatable, of good digestibility and a valuable protein source in the manufacture of prepared foods.

VEGETABLES

Vegetables can be classified into three kinds when considering their use as foods. First, those where the whole plant or leaves and stem are used. These are the green vegetables like lettuce, cabbage, Brussels sprouts, cauliflower, which may be eaten raw or cooked. They have a high water content, a high fibre content and although they are important in the human diet, are not of much value to cats and dogs. They are not usually very palatable to these species and their bulk and indigestible fibre content mean that dogs and cats would have to eat very large amounts to obtain a significant contribution to their nutrient intake. Some dogs and fewer cats may eat cooked vegetables but they do not usually make much contribution to the total diet. Dogs but not cats may obtain some vitamin A from this source. Vegetables are good sources of B vitamins but these may be destroyed in cooking or lost in cooking liquors if these are discarded.

The second class of vegetables are roots or tubers, the main storage organ of the plants, which are rich in starch. Examples are potatoes, carrots and turnips. Root vegetables are not often given raw to dogs and cats because they are so poorly digested in these species. Cooking gelatinizes the starch and makes it more digestible and most dogs will eat cooked potatoes or carrots. Their main nutritional value is as a source of energy, although carrots would also provide some vitamin A for dogs. Cats do not commonly eat vegetables because of their relative lack of palatability. However, there is no risk in including these materials in the diets of cats and dogs.

The third class of vegetables are those where the seed is eaten. This class includes peas and beans. They are relatively rich in protein and provide more energy than stem or root vegetables, other than potatoes. They are fairly good sources of most B vitamins. Green peas, broad beans and runner beans are acceptable to most dogs when cooked but rarely form a major part of the diet. Soya beans are a special case. They are a major source of protein and energy for humans in many parts of the world and are used world-wide for animal feeding, either as whole beans or more commonly after processing to remove the oil. Soya beans have a tough outer seed coat or hull which is removed mechanically before extracting the oil by a combination of grinding and treating with solvents. The residue which is left contains the protein, carbohydrate and mineral portions of the seed with some small amounts of oil. It may be 'toasted' or heated to inactivate certain anti-nutritive factors (trypsin inhibitors and haemaglutinins) contained in the seed. These substances are sensitive to heat and the amount of heat treatment is controlled so that it does not at the same time denature the soya protein and reduce its nutritive value. The heat-treated or toasted de-fatted soya bean meal usually has a protein content of 48–50%, 30% carbohydrate, mainly as sugars not starch, 1–2% fat, about 5–5% minerals with 3–5% crude fibre. The protein is of good quality and has high levels of essential amino-acids.

De-fatted soya bean meal may be used directly as a food ingredient or it may be further processed to make textured vegetable protein, often abbreviated to TVP. Most legumes contain complex carbohydrates and simpler sugars which are resistant to digestion by the digestive enzymes of animals like the dog, cat or man. They pass undigested into the large intestine where they may be fermented by bacteria with the consequent production of flatus or intestinal gas. This problem of flatulence is common with other legumes like green peas and beans discussed earlier. The degree to which flatulence occurs in animals fed soya products or peas and beans seems to depend on the amounts fed and on the susceptibility of the animal which in turn probably depends on the bacterial flora present in the gut. It should not be overlooked nor overemphasized when considering the role of these materials as food.

PREPARED FOODS

The manufacture of foods specially prepared for cats and dogs has developed into a large industry over the last 30 years or so. Most people who have a cat or a dog do not have much knowledge of nutrition, much experience of

TABLE 7

Classification of prepared pet foods by water content and method of preservation

Food type	Water content (%)	Preservation technology
Dry	5–12	Drying
Semi-moist	15–50	Reduced water activity by the use of humectants, mould inhibitors, low pH.
Canned foods	72–85	Heat sterilization
Frozen	60–80	Freezing
Chub	70–85	Heat treatment and/or preservatives
Plastic pot	75–85	Heat sterilization

feeding animals, access to cheap supplies of traditional food items, the time or desire to pursue complicated preparation and cooking of food specially for their pet. They depend therefore on reliable prepared foods for a large part of the diet.

Seven out of ten people who own dogs in Western Europe feed canned pet food at least once a week and five out of ten feed it every day. The situation is similar with cat owners. Canned pet food, then, makes a very important contribution to the diet of the Nation's pets, many of which are fed almost exclusively on such foods.

Prepared foods are available in several forms and may be considered in relation to their intended role in the diet or they can also be conveniently classified by their water content and method of preservation as indicated in Table 7. Prepared pet foods may also be classified on their nutrient content, i.e. if they are sold as a complete food, a complementary food, a mixer biscuit, a snack product or a treat. Complete foods contain all the nutrients required by the dog or cat for that particular stage of the lifecycle for which they are sold, e.g. for adults, for puppies. They require no supplementation, except that clean fresh water should always be made available. Complementary foods are not intended for use as the only food on the diet. They may be rich in some nutrients but inadequate in others. Mixer biscuits, as their name implies, are to be fed with other ingredients in the diet, either prepared or fresh foods. They are high in energy and are good sources of certain minerals and vitamins but this will depend on the recipe used by the manufacturer. Their major role in the diet is to provide a relatively cheap source of energy and as such they usually complement high-protein canned meats.

CANNED FOODS

Canned pet foods have been available for many years as a variety of meat- and fish-based products or meat, fish and cereal products. Their properties are well known, they are a very reliable, safe and convenient way of providing moist attractive foods which are highly palatable to the cat or dog. The most palatable are those which contain little or no cereal or carbohydrate source and are presented as meaty or fishy chunks in gravy or jelly. Those which contain significant amounts of cereals are best described as loaf products, meat and cereal or fish and cereal foods.

Because the nutritional needs of dogs and cats are different, manufacturers usually make foods specifically for dogs or for cats; very few are intended for both species. In practice it is quite safe to feed dogs on cat foods, but it may not be safe to feed cats on dog foods and the practice is not recommended. There is adequate choice of brands and varieties to satisfy the needs of most pets within the species-specific range.

Most canned cat foods are formulated to supply a balanced diet, containing adequate amounts in relation to the energy content of all the minerals, vitamins, fats and amino-acids known to be needed by the cat. So if the cat eats enough food to satisfy its energy needs, then it should at the same time obtain sufficient of all other nutrients. When energy needs are high, as during growth or suckling of young, then specially-made high-energy foods may be more appropriate than ordinary foods which are principally intended for adult cat maintenance.

The digestibility of canned foods is very good and hence the nutrient content has a high availability. The soft-moist, meaty or fishy nature ensures good palatability. Canned dog foods too are usually formulated to provide a balanced diet with adequate amounts of all nutrients in relation to the energy content. It is possible to feed dogs satisfactorily on these only, but because the energy content is relatively low, large amounts are needed. This is rather wasteful of protein and also uneconomic. Most canned dog foods are not made with this in mind but are intended to be highly palatable sources of good quality proteins, vitamins and minerals to be fed in conjunction with cheaper biscuits or other mixers which primarily supply energy with some minerals and vitamins. Feeding recommendations of canned foods which are expected to be fed in this manner usually indicate that the proportions of canned food to biscuit mixer can be varied quite widely to provide a more or less palatable energy-dense mixture to suit the needs of particular feeding situations. For most adult dogs an equal volume mixture of canned food and reputable mixer biscuits will provide a highly palatable, nutritious meal.

Canned foods are safe products with a very long storage life, not requiring special storage conditions. They are usually produced by chopping and mixing the main ingredients, adding gravy and processing in a sealed can. The processing involves combinations of time, temperature and pressure of steam which vary with can size and heat transfer characteristics of the recipe, but which are sufficient to kill even the most harmful bacteria. There is little damage to or loss of nutrients from the food except for thiamin which is particularly sensitive to heat and so compensatory amounts are added to maintain adequate post-process levels.

The main ingredients of canned foods for both species are meat, meat by-products, other protein concentrates like vegetable protein, mineral and vitamin supplements and cereals. They provide a consistent and reliable form of feeding. Typical values for the nutrient content of canned foods and biscuit mixtures are shown in Table 8.

SEMI-MOIST FOODS

Meaty types of dog and cat foods with water contents between 15 and 30% can be preserved with a shelf-life of several months by maintaining a reduced water activity. Water activity is a measure of the water which is available for bacterial or fungal growth in or on the surface of a food. These organisms cannot grow and spoil food in dry products because there is not enough water present. Water activity (A_W) is measured as relative humidity at equilibrum and most bacteria will not grow at levels below 0.83 and yeasts and moulds below 0.6. The low water activity in semi-moist foods is achieved by the inclusion in the recipes of humectants such as sugars, salt, propylene glycol or glycerol which 'tie-up' the water. Further protection is provided by the use of preservatives such as sorbates to prevent yeast and mould growth or by the reduction of pH (increased acidity) with organic acids. These foods can be made with a variety of ingredients including meat, meat by-products, soya or other vegetable–protein concentrates, cereals, fats and sugars. The technology allows the water content to vary over a wide range and so the product form may be as fairly dry material (15% water) not dissimilar to a dry food or a soft-moist substance similar in appearance to mince or cubed meat (25–30% water).

The most popular forms presently available in the United Kingdom contain about 25% water and so have a fairly high nutrient density. They are of average or above average digestibility, 80–85% for most nutrients. They do not usually have a strong odour, do not dry up rapidly if exposed to the atmosphere and so can be left in the feeding bowl without becoming unattractive to pet and to owner. Cat products have not been so successful as those designed for dogs and this may be because the cat is more selective in its choice of foods. Some cat products are available and it is likely that palatability problems will be overcome and they will constitute an alternative feeding form for cats in future. Typical nutrient contents of semi-moist foods are presented in Table 8.

DRY FOODS

Dry foods are available for both cats and dogs. Dry dog foods are sold as baked biscuits, extruded and expanded biscuits, or as mixtures of meals and flakes. They may be complete foods or formulated as mixers intended for feeding as part of the diet with protein-rich foods such as fresh meats, fish or canned dog foods.

Mixers are usually based on cereals and contain very little, if any protein

TABLE 8

Typical nutrient content of prepared foods for dogs and cats

Food type	Moisture %	Protein %	Fat %	Ash %	CHO %	Ca %	P %	ME kcal/100g	Protein calories %
DOGS									
Wet food									
Meat in jelly	81	9	5	2	2	0.3	0.3	81	39
Meaty chunks in jelly	79	7	4	3	7	0.5	0.4	83	30
Meat & cereal	74	8	2	2	14	0.5	0.4	94	30
Meat in jelly for puppies	76	10	8	2	4	0.4	0.4	117	30
Semi-moist food	22	20	9	7	42	1.2	1.0	294	24
Dry Food									
Complete	6	22	7	7	58	1.8	1.4	340	23
Mixer biscuit	7	12	10	7	64	1.3	1.1	351	12
Mixture									
Canned meaty dog food & biscuit (3:1 by wt, equal volumes)	63	10	6	3	18	0.6	0.5	149	23
CATS									
Wet food									
Meat in jelly	82	9	5	2	0	0.3	0.3	69	47
Meaty chunks in jelly	82	7	4	2	5	0.4	0.3	68	37
Meat in jelly for kittens	80	11	7	2	0	0.3	0.2	92	44
Semi-moist food	25	28	15	6	26	1.0	1.8	321	24
Complete dry food	6	30	7	7	50	1.2	1.2	312	36

Note: There are many brands of foods available and their nutrient content varies widely. The tabulated values are indicative of the sort of values found. For information about particular branded foods contact the manufacturer at the address given on the label or packet.

concentrates. They may or may not be supplemented with minerals and vitamins to provide a complete balanced diet when fed in appropriate amounts with cooked or canned meats. Many of the cheaper kinds are not so supplemented and so would require other foods and supplements besides meat to provide an adequate diet. Such mixers are little more than cooked cereal with fat added sufficient for baking or extruding. Good quality mixers are supplemented with extra calcium, phosphorus, trace minerals and vitamins to balance their energy content. When mixed with good quality canned meats they provide adequate amounts of all nutrients.

Dry complete foods for dogs come in similar physical forms but differ in their ingredient content. They are usually formulated to supply adequate amounts of all known nutrients for the stage of life for which they are intended. Loss of nutrients particularly of vitamins is limited because the baking or extrusion processes do not require excessive temperatures or time and sufficient supplements are added to counterbalance processing and storage losses.

Because they are dry and do not contain enough water for bacterial or fungal growth, they have a long shelf-life and will keep in dry, cool storage conditions for several months. Complete dry foods are usually made from cereals and cereal by-products; protein concentrates of animal or vegetable origin like for example, meat and bone meal, fish meal or soya bean meal; fats and mineral and vitamin supplements.

Dry cat foods are mostly available as extruded, expanded biscuits formulated to provide, with water, a complete diet for cats of all ages or for adult cat maintenance. Feeding recommendations often suggest that they are fed only as part of the diet with milk, fresh meat or canned cat food. The ingredients of dry cat foods are similar to those of dry dog foods but more emphasis has to be given to the inclusion of proteins and fats of animal origin and some even include fresh meat rather than meat meals. Cats appear to have a need for greater levels of protein than dogs and so protein levels in dry cat foods are frequently higher than in dog foods. Typical analyses of the nutrient content of dry foods are given in Table 8.

Dry foods contain a greater concentration of nutrients and energy per unit weight than foods of higher moisture content and so relatively small amounts are needed to provide a particular quantity of nutrients. Unless they contain large amounts of fibre, the digestibility of dry foods is usually acceptable but often lower than that of meats and canned foods and is similar to cereals. Dry foods are considerably better digested by dogs than cats. Baked or extruded biscuits have been partially or wholly cooked and so are a good source of energy for both dogs and cats. They are easy to store and to dispense.

The main disadvantage of dry foods is that they are much less palatable than moister foods like meat or canned foods. There are considerable variations in palatability between brands of food because manufacturers take considerable trouble to enhance the acceptance of their own particular products. The mixer biscuits for dogs are meant to be fed with canned meat or meat and gravy and their low palatability is less of a drawback than it is with complete foods. Cats seem prepared to accept the crunchy extruded biscuit forms very well but are not usually impressed by meals or flaked foods. Dry foods provide a relatively cheap and useful source of energy which can add considerable flexibility to feeding programmes.

There are many different foods and ingredients which can be used as whole or part of a dog or cat's diet. They all differ in nutrient content and palatability which determine their suitability for inclusion in the pet diet.

BIBLIOGRAPHY

Anon (1977) *Manual of Nutrition,* Eighth edition. Ministry of Agriculture, Fisheries & Food, HMSO London.

Fonnsbeck, P. V., Harris, L. W. and Kearl, L C. (Editors) (1976) First International symposium 'Feed Composition, Animal Nutrient Requirements, and Computerisation of Diets'. Utah Agriculture Experimental Station, Utah State University, Logan, Utah, U.S.A.

Harris, R. S. and Karmas, E. (Editors) (1975) *Nutritional Evaluation of Food Processing,* Second edition. AVI Publishing, Connecticut, U.S.A.

Kendall, P. T. (1981) Comparative evaluation of apparent digestibility in dogs and cats. *Proc. Nutr. Soc.* **40**, 45A.

Liener, I. E. (Editor) (1980) *Toxic Constituents of Plant Foodstuffs.* Academic Press, New York.

Paul, A. A. and Southgate, D. A. T. (1978) *McCance and Widdowson's 'The Composition of Foods',* Fourth edition. HMSO London; Elsevier/North Holland Biomedical Press, Amsterdam, New York.

CHAPTER 5

Feeding Throughout Life

ANNA L. RAINBIRD

Chapter 2 has already detailed the increase in nutrient requirements of the dog and cat during pregnancy, lactation, growth and exercise. Chapter 4 has discussed in detail the wide variety of foods which are available and the characteristics which make them more or less suitable for inclusion in the diet of a dog or cat. This chapter aims to provide a practical guide to the feeding of dogs and cats during different stages of their lifecycle. Although nutrient supply is paramount, there are many other factors to consider. Puppies, kittens, bitches and queens eat food, rather than ingest nutrients and the characteristics of that food can determine its suitability — is it easy for puppies to eat? do they digest it properly? is it palatable enough to encourage adequate intake? Once these questions have been answered and a food decided upon, there are decisions to be made on the total quantities of food, number of meals and the timing of feeding. Each of these different aspects will be considered as they apply to each stage of the lifecycle.

FACTORS TO BE CONSIDERED IN CHOOSING A FEEDING REGIMEN

The amount of food eaten must supply the animal's needs for energy and nutrients if it is to remain healthy. Most animals will eat sufficient of a palatable diet to meet their energy needs, provided that it is within their physical capacity to do so. A satisfactory feeding regimen is one which provides a balanced diet of sufficient concentration that a dog or cat can obtain its daily needs for energy and nutrients by eating an amount well within the limits set by appetite. Such a regimen could consist of a single suitable food or a mixture of foods of different energy and nutrient concentrations.

The daily needs are governed by the physiological status of the animal. For example, young growing animals or lactating females require very much more in relation to their size than normal adults in the maintenance state. The more concentrated and more palatable foods are therefore most suitable for these stages of life where demands for nutrients are high, and intakes may be limited by the bulkiness of some foods.

In general it is not absolutely necessary that the daily intake of nutrients and energy should exactly match requirement. Animals are resilient and able to store some surplus nutrients in body tissues when supply exceeds needs and to use these stores when food or nutrient supply is inadequate. Some nutrients cannot be stored except in very small amounts, examples are water soluble vitamins and amino-acids.

In young growing animals or lactating females, because of their higher nutrient demand, the consequences of under-nutrition or of an improperly balanced diet are seen more rapidly than in adults at maintenance. However, even with adults the aim should be to supply foods which on average meet their needs each day, and certainly do so over a 7–10 day period.

FREQUENCY OF FEEDING

Most pet animals are adults which live mostly indoors in a temperate climate. They are usually not pregnant or lactating, not involved in regular heavy work or excessive exercise and are not subjected to extreme environmental conditions.

Dogs in this situation will usually have a good enough appetite to eat all they require in one meal per day and it is quite satisfactory to adopt a once-a-day feeding regimen. The advantages of a single meal are that it can usually be of sufficient size to satiate appetite; it is more controllable in that errors of dispensing occur only once a day, so under- or overfeeding is less likely; it can easily be fitted into the household routine at whatever is the most convenient time. It is usually best to avoid late evening meals since dogs may need to excrete faeces or urine within a few hours of feeding and this can be inconvenient in the middle of the night.

There is no disadvantage in feeding more frequently than once a day, provided that the total daily intake is limited to the dog's daily needs. Feeding two or three times a day at the same time as family meals is a common practice. The risk is that more food will be given and that overeating and obesity will be the result. The correct number of meals for any adult dog is that which the owner and dog find most convenient. It is desirable to establish a routine and stick to it. Mealtimes are the high spot of most days and dogs quickly become accustomed to being fed at the same time and place each day.

Dogs which are unwell or have poor appetites, or very old dogs, may benefit from being fed two or more times daily with smaller meals. Very young growing puppies should be fed 4 to 5 meals per day but once weaned this can be decreased to 3 to 4 meals a day. Once the puppy has reached half of its adult weight the number of meals can be further decreased to 2 per day if sufficiently concentrated foods are given. For maximum benefit the meals should be spread as far apart throughout the day as possible.

Working dogs and lactating females which require energy and nutrient intakes of 2 to 4 times that of a normal adult dog in relation to their size will also benefit from more frequent meals since this gives them a better opportunity to ingest the large amounts of food they need. There is no single optimum number of feeds per day, except that which is convenient for the owner and which provides adequate opportunity for energy and nutrient requirements to be met.

Cats are rather different. It is generally thought that cats which have gone back to the wild, or genuine wildcats, must be opportunist feeders eating as and when they catch and kill prey. It might therefore be thought that they could be adapted to large meals at irregular intervals. However, it is likely that small rodents and birds which would constitute the major part of their diet are not big enough for one kill to provide for the animal's daily needs and more than one meal per day would be more usual.

Observations on domestic cats given completely free access to palatable foods have demonstrated that cats prefer to eat many small meals (12–20 meals per 24 hours) rather than one or two large ones. Most cats appear to be able to regulate energy intake so that they do not overeat and become obese when food is made available in this way. It therefore seems reasonable to feed cats several small meals rather than one or two large ones. This is quite satisfactory in practice and many pet cats are fed 'on demand' with food made available whenever the animal asks for it throughout the day. The average pet cat in Europe weighs almost 4 kg and needs about 320 kcal metabolizable energy per day. This would be provided by approximately 100 g of dry food or one large can of a typical canned cat food. Very few cats are prepared to eat this amount of food in one meal and so at least two or more meals per day are necessary. It is therefore good husbandry to feed cats to appetite several times daily. Alternatively cats may be given free access to food which is renewed at least twice daily.

CATS AND DOGS ARE INDIVIDUALS

Most cats and dogs are kept in households in which they are the only cat or dog, or where there is not more than one other animal of the same species. Pet owners therefore tend to treat them as individuals and need to develop feeding practices suitable to their animal; taking account of the particular circumstances, likes and dislikes of the animal and of their own view of convenience, cost, variety and suitability of foods. They need to identify the particular needs of their animals and find a combination of food or foods to meet them.

It is not possible in a book of this kind to prescribe particular dietary regimens for each pet animal in every environment. Therefore the feeding guides in this chapter are intended only as GUIDES to the use of prepared foods for average dogs and cats in the usual range of environments found in Western households. It is relatively simple for the individual owner to use these guides as a starting point to obtain an approximate estimate of their pet's needs, then by observation of the animal to decide whether to feed more or less, and by substitution of one food for another to arrive at a suitable regimen.

The feeding guides found on the packets, cans or other packaging of manufactured foods are usually intended to apply to the average adult dog or cat living mostly indoors and given a moderate amount of exercise.

ENERGY REQUIREMENTS OF DOGS AND CATS

The range in bodyweights of different breeds of cats is quite small. This makes calculating their energy requirement relatively easy compared with the dog. Most adult cats require between 70–90 kcal metabolizable energy per kg bodyweight per day. Lethargic cats taking little exercise will have lower energy requirements than adventurous cats which take long trips outside in all weathers. Cats are very good at regulating their own food and energy intake and will usually compensate for changes in their requirements.

Adult dogs in peak condition vary in size from around 1.5 kg up to 118 kg. Metabolizable energy requirements of dogs in kcal/day can be estimated by using the formula $100W^{0.88}$ (NRC, 1985) where W is the bodyweight of the dog measured in kilograms. The term $W^{0.88}$ is the metabolic body size of the dog. The aim of using an estimate of metabolic body size (bodyweight in kg raised to a certain power function) is to make an allowance for the wide range of bodyweight in dogs and the fact that energy needs are not directly proportional to body weight but to the weight of actively metabolizing tissue.

A power function of 0.75 was shown by Brody (1945) to give a reasonable estimate of the metabolic body size of a wide range of animals from mouse to elephant but is no more than an approximation. From all data available on the energy requirements of adult dogs, Thonney (1985) calculated that the best power function to use for the dog species was 0.88. Having calculated approximate energy needs of adults using the formula $100W^{0.88}$ the amounts of food needed to meet these needs can be calculated from a knowledge of the energy value of the foods. These average amounts can then be rounded to a convenient unit for feeding, which for canned foods may be half or a quarter can and for dry or semi-moist foods 10 or 20 g units. Since very few people actually weigh food for their pets but are more likely to feed it by some form of volume measurement or by eye, it is common to find feeding recommendations given as cupfuls for dry and semi-moist foods.

In order to allow for variability between dogs and for differences in environment, actual feeding recommendations are often quoted as a range which may cover from 20 or 25% below to 20 or 25% more than the estimated average. Although this may seem a very approximate method of arriving at feeding recommendations, it does in fact work out very well in practice. It should not however surprise owners to find that their particular dog or cat needs much less than the amounts suggested, particularly if additional snacks, treats or table scraps are being added to the diet, or in the case of cats, if they are eating the odd mouse. The best criterion for judging the adequacy of a feeding regimen is the health and appearance of the dog or cat. If they are in good condition with good skin and coat, alert, active and neither too fat nor thin, they are on an adequate regimen.

FEEDING KITTENS

Kittens at birth usually weigh between 80 and 140 g with most weighing around 100 to 120 g. They are entirely dependent on the milk supply of the queen for their nourishment for about the first 4 weeks of life. At 4 weeks of age the kittens will usually have trebled their birthweight and begin to explore their surroundings. At this time the process of weaning or gradual replacement of the queen's milk by other foods, can begin. Weaning is a time of learning and is best done gradually to avoid upsets to the digestive system. The kitten has to become accustomed to new tastes and textures of foods and its digestive system has to learn to cope with new kinds of proteins, fats and carbohydrates. It is often suggested that the first new food should be milk-based. Although most kittens will drink such foods, their use is not essential. Given the opportunity kittens will begin to eat the queen's food at around 4 weeks of age. They will also start to eat finely minced or chopped moist food, if this is provided in a shallow dish for easy access. Because they will only eat small amounts to begin with, it it best to use highly palatable, moist, meaty foods. Some canned foods are specifically made for kittens and have a higher energy and nutrient density than most other products and are thus the most suitable type of prepared food to use.

An ideal weaning food would be of a similiar energy density, to queen's milk. At weaning the kitten requires approximately 260 kcal/kg bodyweight, compared with 70–90 kcal/kg bodyweight required by a fully grown cat; therefore its energy intake, in relation to bodyweight is three to four times that of an adult cat. In addition to being concentrated, food for kittens must have a high digestibility and availability of nutrients to allow maximum benefit to be gained from a reasonable level of intake. Further, it needs to have a smell and taste which encourages the kitten to eat and a texture which makes this possible. Dry and semi-moist foods are concentrated sources of nutrients and may be suitable for very young kittens. Some kittens may find dry foods difficult to eat unless chopped or soaked. The digestibility of dry foods is usually lower than that of canned food. This may make them less suitable for weaning kittens unless they have been specifically formulated. Poor nutrition of kittens can therefore be the result of either inadequate intake of food which may be nutritionally satisfactory but which is indigestible, insufficiently palatable or too diluted. Unfortunately, the initial symptoms of a single nutrient imbalance (inappetence, poor growth) are also those associated with an inadequate level of intake for these reasons.

Most queens will continue to suckle their kittens until 7 or 8 weeks after parturition. However by this time the proportion of the kitten's total nutrient intake coming from supplementary food should be at least 80–90% and they

TABLE 9

Estimated energy requirements of young growing cats in relation to their age and expected body weight

Age (weeks)	Expected body weight (kg)*	Average metabolizable energy requirement (kcal/kg/day)	(kcal/day)
8	0.6–1.0	260	160–225
10	0.8–1.2		175–280
12	1.0–1.4		222–275
14	1.2–1.6	200	235–300
16	1.4–1.9		246–325
18	1.6–2.2		274–355
20	1.8–2.5	150	280–375
24	2.1–2.8		269–363
28	2.3–3.1		288–395
32	2.5–3.4		301–403
36	2.7–3.7		313–426
40	2.9–4.0	100	316–436
44	3.0–4.1		300–401
48	3.0–4.2		257–364
52	3.0–4.3	80	240–344

* Lower weight for females, higher weight for males.
Data taken from feeding trials conducted at the Waltham Centre for Pet Nutrition

can be finally separated from their mother. Thereafter the kittens are fed independently of their mother.

A weaned kitten at 8 weeks of age may weigh anything from 600 g to 1 kg. By this age the males are already significantly heavier than the females (Loveridge, 1987). A kitten will be very active and spend much of its waking time in play. The energy requirements of growing kittens are variable depending on size, activity and environment. At weaning a kitten will require approximately 200 kcal metabolizable energy per kg bodyweight. Kittens grow very rapidly if fed a palatable, balanced diet to appetite. They will achieve a nearly adult weight of around 2.4 kg for females and 3.2 kg for males by 6 months of age. At this stage they are still very active and will usually eat enough food to supply 150 kcal/kg bodyweight per day. It is not until they are nearly a year old that they settle down to the average adult energy intake of 70–90 kcal/kg bodyweight. Table 9 is a guide to the estimated energy needs of growing kittens of various ages and weights. The weights for each age group are typical of those found in the cat population as a whole. The estimates of energy requirement are based on the voluntary food intake of growing kittens fed a variety of diets (Loveridge, 1987 and Waltham Centre for Pet Nutrition data). In practice during the period of growth, kittens should be fed as much as they will eat as they regulate food intake well. However, a close eye must be kept on their condition to ensure that they are making skeletal and muscular growth and are not becoming obese.

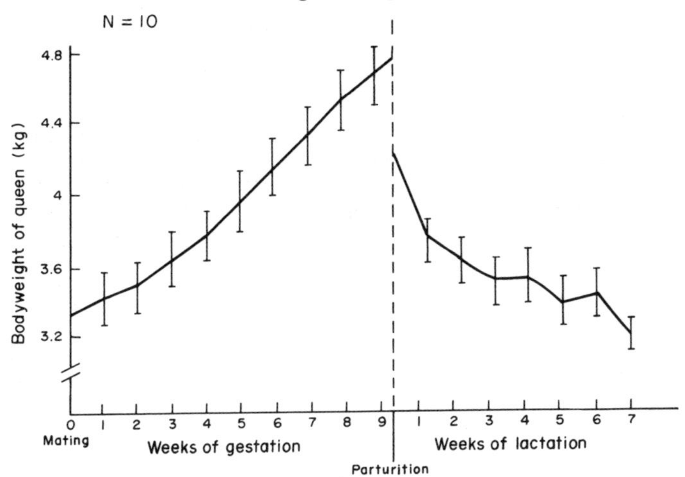

FIG. 12. Bodyweight changes of queens during gestation and lactation (Loveridge, 1986).

FEEDING ADULT CATS

A survey has shown that the average weight of adult cats in the United Kingdom is 4 kg. Although most cats weigh in the 3–5 kg range, smaller and larger cats are not uncommon. Male cats can reach 6–8 kg without being obese. It is relatively easy to feed cats well by feeding them to appetite (giving them as much food as they want to eat at each meal) several times a day. Cats can become very particular about their food. Therefore it is a good idea to introduce a variety of different flavours or food types (canned, semi-moist, dry or fresh) into their diet whilst they are still growing. This will make it more likely that they will readily accept a variety of foods as adults. If prepared pet foods are to be used, only ones of good quality from a reputable manufacturer should be chosen. Most cats with access outdoors will probably supplement their diet with small rodents and birds and many will be given small amounts of human foods such as milk and meaty or fatty scraps. These 'extras' can account for between 10–20% of cats' daily needs. There is no cause for concern if a cat does become fixed on one variety or brand of prepared food since cat foods manufactured by reputable companies are formulated to be nutritionally complete. If the cat only eats one type of fresh food, care must be taken to ensure that it is adequately supplemented with minerals and vitamins to give a balanced diet. Provided the cat is healthy and not obese and remains so, the diet is satisfactory.

FEEDING CATS IN PREGNANCY AND LACTATION

The extra nutrient requirements of the queen during pregnancy and lactation have already been discussed in Chapter 2.

Figure 12 shows the mean bodyweight of queens during pregnancy and

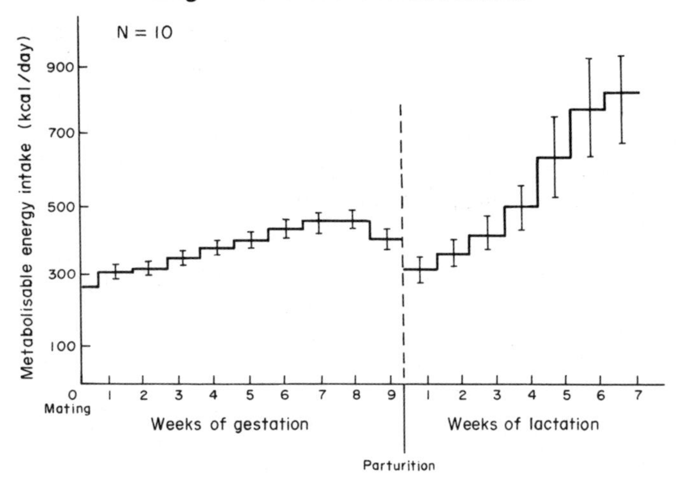

FIG. 13. Energy intake of queens during gestation and lactation (Loveridge, 1986).

lactation (Loveridge, 1986). The queen's bodyweight begins to increase steadily from 1 week after mating until parturition. This is in contrast to the dog where most bodyweight gain occurs in the last third of pregnancy. The weight of the queen increases by approximately 40% during the 9th week of pregnancy. At parturition, only 40% of the weight gained by the queen in pregnancy is lost. During the period of lactation the queen gradually loses weight until, at weaning 8 weeks after parturition, she is back to her mating weight. The number of kittens that the queen is carrying during gestation and rearing in lactation has a considerable effect on the weight gained in pregnancy and the weight reduction in lactation (Loveridge, 1986). There is usually no need to provide any special feeding for cats which are pregnant, particularly if a regime of feeding to appetite with a variety of foods is being followed. The energy requirement of the queen increases during pregnancy as body tissue is laid down. Therefore more food should be offered to the queen to allow her to satisfy her appetite. The extra nutritional needs of pregnancy are small and will be adequately catered for by more of a balanced diet, whether it is a prepared food or mixed diet.

Once the kittens have been born, the nutritional needs of the queen increase rapidly because she has to provide, through her milk, the nutrients and energy needed for the very rapid growth of her offspring until they begin to eat supplementary foods when about 4 weeks old.

The energy requirement of the queen in lactation is dependent on the number of kittens in the litter and the amount of milk produced. Queen's milk has an energy density of 106 kcal/100 g which is very high compared with cow's milk at 65 kcal/100 g (Baines, 1981). The energy requirement of the queen during lactation can reach three to four times her normal maintenance needs (Fig. 13).

The queen should therefore be encouraged to increase her energy and nutrient intake by the provision of frequent meals and by offering more concentrated foods. A mixture of prepared and fresh foods, including milk, may tempt the appetite of the queen allowing her to satisfy her needs. Water requirements also increase in lactation because of that lost in the milk. A fresh supply of water should always be available to all cats whatever they are fed on.

FEEDING PUPPIES AND GROWING DOGS

Feeding the puppy begins at weaning and although this can be started and is regularly done by some breeders by placing small amounts of food directly into the puppies' mouths before they are up and walking, it is a laborious business and any practical advantages are minimal. Weaning will begin naturally when puppies are 3–4 weeks old and actively exploring their surroundings. At this age they will readily take to soft, wet foods which are easy to ingest. Although many people assume that milk and milky feeds should figure prominently in a weaning regime, they are not essential. There are many equally suitable alternatives available.

Like kittens, puppies will eat minced or chopped moist food. The litter should be fed together from one or two trays depending on the litter size to encourage competition between the puppies. Highly palatable, energy-and nutrient-dense foods are those most suitable for weaning. There are some prepared foods specially formulated for puppies. Canned foods are very acceptable to very young puppies during weaning but dry foods will need moistening with milk or water in the early stages and are often lower in palatability.

In the early stages of weaning (3 to 4 weeks after parturition) the bitch's milk is still the most important source of nutrients and the puppies' digestive and immune systems are learning to handle new sources of nutrients. Thereafter the intake of other foods quickly increases and most litters can be completely weaned on to a varied diet or a single complete food by the time they are 6 weeks old. Once weaned, puppies grow at a rapid rate and need to ingest very large amounts of energy and nutrients in relation to their size. Like kittens, in general puppies require two or three times the energy intake of an adult of the same bodyweight.

It is therefore necessary to use more concentrated foods. At 4 weeks of age four meals a day should be offered. This can gradually be reduced to one or two meals a day as the animal approaches its adult bodyweight. Because of the wide range of bodyweights of dogs and ages at which they achieve their final mature bodyweight (Fig. 14) the timing of changes in feeding frequency and amounts of food will depend on the breed of dog. Small and toy breeds of dog reach their adult bodyweight at 6–9 months of age, whereas giant breeds of dog are not mature until they are 18–24 months old.

It is impossible to give advice on the exact amount of food and thereby

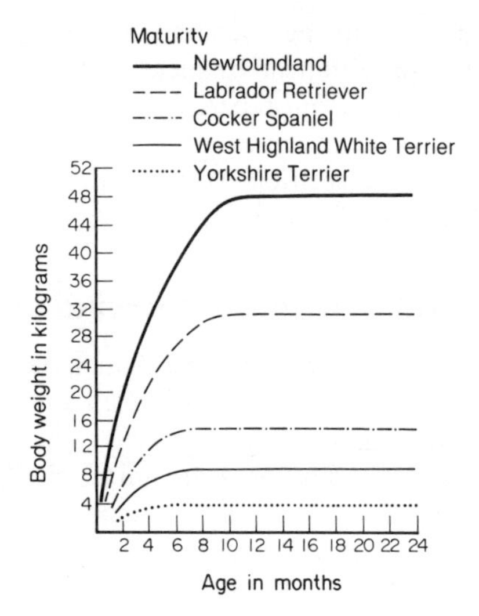

FIG. 14. Average weight of various dog breeds.

energy, which should be fed at each occasion to an individual dog. Besides differences in breed and stages of life there are also differences in activity and environmental temperature. Unfortunately established standards of size and weight for growing dogs of many breeds are not available. Ideally they should also be divided according to sex. A puppy should put on weight and grow at a rate which is neither too slow nor too fast. The weight, size and state of nutrition should be compared with litter mates or other dogs of the same age and breed.

There has been some debate about whether or not puppies should be fed to give the maximum possible rate of weight gain of which they are capable. From all the evidence available to date, it appears that dogs with optimum qualities of conformation are usually reared at a moderate rate of growth so that development to the usual adult size and conformation takes place over a slightly longer period. Controlled feeding of a balanced diet facilitates optimal skeletal characteristics, rather than maximal rate of growth.

Puppies of toy and small sized breeds will grow satisfactorily and develop to the usual adult size if fed about 260 kcal/kg metabolic bodyweight ($kg^{0.75}$) per day from weaning until half grown. Once half grown, the energy allowance can be decreased to about 200 kcal/kg metabolic bodyweight per day, gradually being reduced towards the adult requirement of 100 $kcal/kg^{0.88}$. Medium and large breeds of puppies will require about 335 $kcal/kg^{0.75}$ from weaning to half grown, this is then reduced to 250 $kcal/kg^{0.75}$ from half grown to adult. Giant breeds like Great Danes and Newfoundlands have been found to require considerably more energy in the first 8 months of life if they are to achieve satisfactory growth (370 $kg^{0.75)}$. This can be reduced to 300 $kg^{0.75}$ until 15 months of age and then gradually reduced further until the adult

TABLE 10
Energy requirements of dogs in kcal/dog/day

Age in months	Toy e.g. Yorkshire Terrier	Small e.g. West Highland White Terrier	Medium e.g. Cocker Spaniel	Large e.g. Boxer, Labrador	Giant e.g. Great Dane
2–4	220–310	310–685	685–925	925–1850	1850–2775
4–6	310–465	465–685	925–1390	1850–2775	3700–4625
6–9	Feed adult food	465–685	1390–1850	1850–2775	3700–4625
9–12		Feed adult food	925–1390	1390–2315	3240–4165
12–18			Feed adult food	1390–1850	2775–3700
18–21				Feed adult food	2775–3700
21 onwards					Feed adult food

requirement is reached. Table 10 presents guidelines of the average daily calorie requirement of each class of dog from weaning to adult.

An important aspect of feeding growing dogs which is of concern to almost all who have ever owned a puppy but which does not receive much attention in the literature is the effects of diet on the amount and consistency of faeces and the frequency of defaecation. It is a very difficult topic to discuss objectively. Faeces are not simply indigestible residues of food but also contain bacteria, mucus, the dead cells from the lining of the gut and materials actively excreted into the lumen of the large bowel. Faecal bulk or volume depends on several things including the amount of indigestible matter in the food, but also the fluid balance within the alimentary tract. Owners usually have to gather and dispose of dog faeces and this job is more easily performed when faeces are well-formed, firm and not excessively fluid; that is more like sausages than scrambled eggs in consistency.

Food is only one of the factors influencing faeces bulk and consistency. In general, foods of high digestibility result in smaller amounts of well-formed faeces but this is not always so. One of the factors influencing fluidity of faeces is the rate of passage of digesta through the tract. The ingestion of very large amounts of food may speed up transit through the gut and give insufficient time for water reabsorption in the large bowel. Also bacterial fermentation of some food residues may produce gas and materials which irritate the gut and cause a decrease in transit time. Anything which upsets the water balance of the gut can result in very wet loose faeces. Dietary factors which regulate the final consistency of faeces have not been clearly identified, so it would be unwise to suggest that the inclusion or omission of certain ingredients will necessarily result in satisfactory faeces. Experience suggests that puppies generally produce softer faeces than adults and that this is related to the large amount of food eaten relative to their size. Feeding several smaller meals of more digestible foods is usually helpful. Some puppies take longer to settle to diet changes, and high fat diets which slow down transit times and are highly digestible may at first result in greasy loose stools if introduced abruptly into the diet, despite usually being satisfactory once puppies have adapted to them. Observation over several days following a change of diet or feeding management is probably the best way of arriving at a suitable combination of food and management for the individual puppy. Dietary management is at least as important as the composition of the food fed.

REARING MOTHERLESS PUPPIES

Rearing motherless puppies is a very demanding task. A great deal of dedication is needed to ensure a satisfactory outcome. Nevertheless, successful rearing of orphaned puppies is very rewarding and engenders a strong sense of achievement.

A puppy's mother does not have to die to deprive it of the vital nourishment it requires. The bitch may reject the puppy or be producing too little milk. In either of these situations further investigation is needed to try to discover why the bitch has no milk or if there is any abnormality of the puppy which might cause its rejection. If there is some obvious deformity present, it would be wiser not to attempt to rear such individuals.

The most obvious alternative to a bitch rearing her puppies is for another individual to do so, that is for another bitch to act as a foster mother. Although this is a very much more satisfactory arrangement than trying to hand-reared puppies, the chances of a bitch at the right stage of lactation, and with sufficient resources to rear a litter being available at just the right time are poor. It can however be done and good communications within a breed club obviously improve the chances of finding a foster mother. Motherless puppies have vital requirements in two main areas — nutrition and provision of a suitable environment. There are two very important aspects of husbandry to consider: the ambient temperature around the puppies and the stimulation of urination and defaecation of each puppy. Newborn puppies are unable to control their body temperature effectively. Ideally the environment would be controlled by means of an incubator. Alternatively a heating pad with adequate insulation of the pen can be used. Newborn puppies are normally either eating or sleeping. After puppies have fed, a vital aspect of tending motherless puppies is to simulate the mother's tongue action on the ano-genital area which provokes reflex defaecation and urination. The application of this stimulus has to be taken over by the person tending the puppies. The necessary result can be achieved by applying a piece of damp cotton wool at the ano-genital area. It is sometimes possible to effect the same response simply by running a dampened forefinger along the abdominal wall. This stimulation should be a routine carried out after each feed. After this procedure each puppy needs to be carefully cleaned. At about the age of 3 weeks, puppies are able to relieve themselves without their mother's stimulation or the simulated equivalent. They will begin to wander away from their

TABLE 11
Average analysis of milk of various species (Baines, 1981)

	Bitch	Cow	Goat	Cat
Moisture %	77.2	87.6	87.0	81.5
Dry matter %	22.8	12.4	13.0	18.5
Protein %	8.1	3.3	3.3	8.1
Fat %	9.8	3.8	4.5	5.1
Ash %	4.9	5.3	6.2	3.5
Lactose %	3.5	4.7	4.0	6.9
Calcium %	0.28	0.12	0.13	0.04
Phosphorus %	0.22	0.10	0.11	0.07
*Energy (kcal/100g)	135	66	70	106

* Calculated using protein 4 kcal/g, fat 9 kcal/g and lactose 4 kcal/g.

bed to relieve themselves, but do not do so in particular places until they are fully weaned at about 7–8 weeks.

As the puppy develops, active periods become more evident between sleeping and feeding. After about 2.5 – 3 weeks, puppies begin to explore their surroundings. A larger area is needed for their increasing activities. Great care must still be taken to avoid chilling, excessive disturbance or exposure to hazards outside the puppy pen. Puppies grow at a rapid rate. They will double their birthweight in a matter of days. Because of this, puppies require quite large quantities of their mother's milk or a food which can substitute for it. The food has to be a concentrated source of nutrients based on the composition of normal bitches' milk. Table 11 above shows the average composition of milk from bitches, cows, goats and cats. It is clear that cow's milk is grossly inadequate as a substitute for rearing puppies. The protein, fat and calcium levels are much lower and the calorie density is only about half that required. The level of lactose present is probably higher than puppies can tolerate for any length of time. Many commercially available bitch's milk substitutes can now be bought. They are usually based on cow's milk which has been modified to resemble bitch's milk more closely. Whatever product is used, feeding by hand is a very time consuming procedure. Foods can be administered by means of a small syringe, a puppy feeding bottle or an intra-gastric tube. Dried milk feeds should be reconstituted daily and fed warm (38°C). Food must be given slowly and must not be forced into the puppy. When feeding from a miniature bottle, the hole in the teat may need to be enlarged so the flow is improved and the puppy does not suck in air.

When puppies begin to start exploring their surroundings, the milk substitute can be made available to them in shallow dishes. From 3 weeks good quality foods may be introduced in the same way as discussed earlier. These can be mixed with the milk substitute to begin with and then offered separately.

FEEDING ADULT DOGS

A healthy adult dog can be fed totally on one food or with any number of combinations. It is not possible to cover all types and combinations of diets but Tables 12 and 13 provide a summary of the average energy needs of adult dogs and amounts of foods required to meet these needs on prepared and mixed food diets.

The tables can only serve as a starting point for a feeding regime, the actual amounts needed for the maintenance of any particular dog have to be obtained by trial and error and careful observation of the dog's health when the amounts given are varied. The suggested proportions of canned meat to biscuits are equal amounts by volume, approximately 3:1 meat:biscuit by weight. This ratio provides adequate amounts of all nutrients and is chosen to incorporate all the palatability benefits of the canned food and the economy of the biscuit.

FEEDING THE BREEDING BITCH

To plan a sensible feeding programme for the bitch, it is necessary to understand what are the extra nutritional and physical demands made by breeding. A normal healthy bitch does not have any large increase in growth of new tissue during the early part of pregnancy. Most foetal growth takes place during the last 3 weeks and although there is considerable development of mammary and uterine tissues before this, the extra need for nutrients and energy over and above maintenance requirements are quite small. A bitch in good condition at mating will not require any special food during pregnancy and can continue to receive her usual balanced diet. All that is necessary is that the amount is gradually increased during the second half of gestation. It has been found that increasing the total food allowance by 10% each week from the 6th week onwards, so that intake at birth is approximately 50% more than at mating, is a satisfactory regime for most dogs.

It may happen that a bitch with a large litter may have such an enlarged abdomen and such reduced activity that her appetite falls during the last week or 10 days of pregnancy. In these cases it is sensible to feed several smaller meals and perhaps introduce or increase the amounts of concentrated foods which will probably be used in lactation anyway. The objective is to have a bitch at parturition which is not overfat and which has maintained her appetite.

Lactation presents the biggest test of nutritional adequacy of any feeding regime. The bitch must eat, digest, absorb and use very large amounts of nutrients to produce sufficient milk of adequate composition to support the growth and development of several puppies. Experience and theory both indicate that the amounts needed are very large. Consider a Labrador bitch of 28 kg with a litter of eight puppies, total litter weight 20 kg at 3–4 weeks of age. At this stage the puppies require about 200 kcal/kg bodyweight per day which they obtain from the bitch's milk. The bitch therefore has to supply

TABLE 12

Average energy needs of healthy adult dogs and amounts of different types of prepared food needed to meet them

Weight of dog (kg)	Typical breed	Energy requirement (kcal/day)	Food to provide this amount of energy: Canned food + mixer 3:1 by weight, approx. equal volumes			
			meat (g/day)	mixer (g/day)	Semi-moist food (g/day)	Complete dry food (g/day)
2	Yorkshire Terrier	115	60	20	38	33
8	Fox Terrier	460	240	80	151	132
14	Cocker Spaniel	806	420	140	264	230
20	Border Collie	1151	599	200	377	329
26	Basset Hound	1496	779	260	491	427
32	Labrador Retriever	1841	959	320	604	526
40	Deerhound	2302	1199	400	755	658
50	Bull Mastiff	2877	1499	500	943	822
70	Irish Wolfhound	4028	2098	699	1321	1151

Energy densities of prepared foods will vary depending on the recipe used. For the purpose of this table energy densities have been assumed as follows:

Canned food	0.7 kcal/g
Mixer	3.5 kcal/g
Semi-moist food	3.05 kcal/g
Complete dry food	3.5 kcal/g

TABLE 13

Average energy needs of healthy adult dogs and amounts of mixed feeding regimes needed to meet them

Weight of dog (kg)	Typical breed	Energy requirement (kcal/day)	Food to provide this amount of energy: Tripe (g/day)	Tripe + Mixer equal weights: tripe (g/day)	Tripe + Mixer equal weights: mixer (g/day)	Mince + Mixer equal weights: mince (g/day)	Mince + Mixer equal weights: mixer (g/day)
2	Yorkshire Terrier	115	128	26	26	21	21
8	Fox Terrier	460	512	105	105	84	84
14	Cocker Spaniel	806	895	183	183	146	146
20	Border Collie	1151	1279	262	262	209	209
26	Basset Hound	1496	1662	340	340	272	272
32	Labrador Retriever	1841	2046	419	419	335	335
40	Deerhound	2302	2558	523	523	419	419
50	Bull Mastiff	2877	3197	654	654	523	523
70	Irish Wolfhound	4028	4476	915	915	732	732

Mineral and vitamin supplements must be added to the diet in the correct portions to ensure that nutritional balance is achieved.

Energy densities of foods will vary depending on the source. For the purpose of this table typical energy densities have been assumed as follows:

Tripe 0.9 kcal/g
Mixer 3.5 kcal/g
Mince 2.0 kcal/g

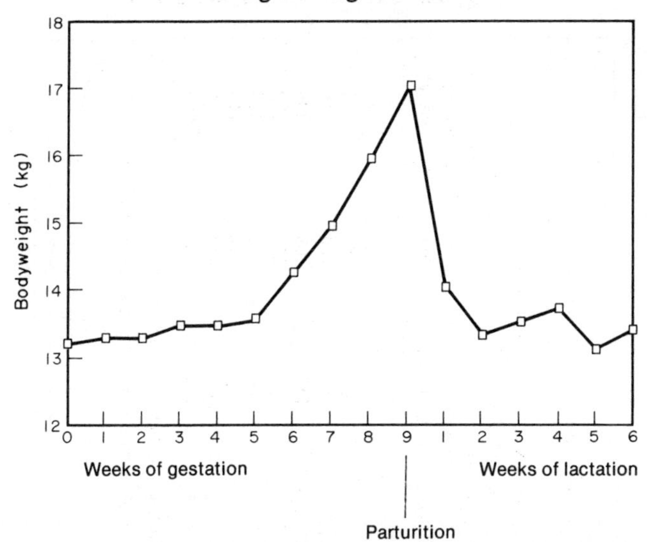

FIG. 15. Typical bodyweight changes of a beagle bitch during pregnancy and lactation.

4000 kcal as milk each day. Bitch's milk contains about 1300 kcal/litre and so the amount of milk needed is at least 3 litres or 5.5 pints. There are obviously some losses of energy in the production of milk by the bitch but if it is assumed that the process has an efficiency of 75% then in order to produce 4000 kcal as milk, the bitch must obtain 4000/0.75 or 5333 kcal from her food. In addition, to maintain her own bodyweight and condition she will need her usual 1800 kcal/day. Her total energy requirement is therefore 7213 kcal or nearly four times her maintenance requirement. Obviously bitches with greater milk production would need even more food. If the bitch is unable to produce enough milk or to eat the amounts of food she needs, then earlier supplementary feeding of puppies may be necessary if they are to do as well as they should.

The above calculation is based on estimates of the energy needs for satisfactory milk production but requirements for other nutrients are similarly increased. Protein quantity and quality will affect milk production. It is therefore necessary to ensure that the extra food supplied is of good quality and is not made up only of high fat or high carbohydrate foods. Because the amounts needed are very large, it is usually necessary to feed several meals per day or *ad libitum*. Figure 15 shows the bodyweight changes of a Beagle bitch during gestation and lactation and Fig. 16 shows the average energy intake. These figures may be compared with Figs. 12 and 13 for the cat.

Many diets are suitable for lactating bitches. If canned food and mixer is the usual diet, it may be necessary to increase the proportion of biscuit used to achieve satisfactory energy intakes. The use of more concentrated foods like puppy food, semi-moist or complete dry foods is another way to achieve higher energy intakes. If the dog is willing to eat different kinds of foods this makes it easier, although it is quite satisfactory to keep to one or two foods if this is the

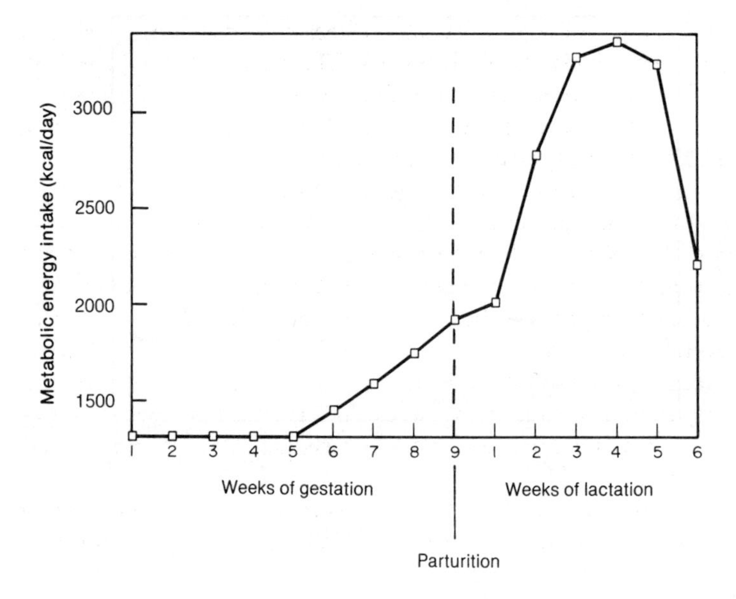

FIG. 16. Typical energy intake of a beagle bitch during pregnancy and lactation.

normal regime. The important point is that it is necessary to encourage the bitch to eat up to four times her normal maintenance requirements. The breeding bitch does not require special vitamin/mineral supplements if a balanced diet is used.

FEEDING WORKING DOGS

The working dog is also an adult but one which has more variable and much greater energy needs. The amount of extra energy required depends on the nature of the work. Working dogs perform many different functions, for example, acting as guide dogs for the blind to pulling sledges in polar regions. Depending on their function, working dogs have very different training and working schedules necessitating different diets and feeding regimes. The energy requirements of the working dog will depend on the environment, the amount of exercise and the type of work undertaken. Working dogs will have rest and training days as well as working days and hence the energy requirements for each of these occasions will differ.

Sheepdogs covering many miles over rough terrain, often in inclement weather, may need considerably more than patrol or guard dogs. They can sometimes need 2–3 times normal maintenance amounts but their feeding is straightforward. Simply give more of the same sort of food or, because the quantities needed may be so large, use more of the concentrated semi-moist or dry foods. If there is a rest period during work it is sensible to give a small meal at that time but reserve the main meal until after work. All working dogs should only receive a small meal prior to working, since it is often inconvenient for guard and patrol dogs to have to defaecate during a duty period. A full stomach is not conducive to efficient work. The main meal should perhaps provide 2/3 of the daily needs and since dogs are often tired and have poor appetites, the more concentrated and palatable foods should be used. Working dogs should be given an opportunity to drink during the working period.

Hard work induces stress, so feeding for hard work should include consideration of nutritional requirements for stress as well as provision of fuels for muscular exercise. The preponderant fuels are fats and soluble carbohydrates, and provision of these nutrients is the main aim of diets designed to sustain or promote hard work. Hard work and stress tend to engender several clinical conditions that may be averted or mitigated by appropriate feeding programmes and diets. These conditions include the diarrhoea–dehydration–stress syndrome, exertional rhabdomyolysis ('tying-up'), gastric dilation-volvulus, lower bowel bleeding, stress anaemia, and metatarsal fractures.

Sledge dogs are unlike any other group of working dogs in the degree of effort they are required to make over long periods in hostile environments. They will require more food and energy not only as a result of exercise, but also to maintain their body temperature. Significant correlations have been drawn between food consumption and ambient temperature. Sledge dogs (Huskies) in the Arctic have been estimated to require 2800–3200 kcal/day at base camp, increasing to in excess of 5000 kcal/day on a working day, i.e. 2 and 3 times their normal adult requirement. A diet for hard-working dogs should be of a high energy density, of high digestibility, of high palatability and nutritionally balanced. A diet of high energy density can be obtained by including a high proportion of fat in the diet. High-fat diets must be balanced with a high concentration of other essential nutrients to match the so-called 'empty' calories of fat. Carbohydrate is not an essential nutrient for the dog and no adverse effects have been observed when feeding a carbohydrate-free diet to working dogs so long as the correct balance of protein, fat, vitamins and minerals has been maintained. If a high protein, high fat diet is to be fed to the working dog, the protein level must be sufficiently high to allow carbohydrate synthesis by the dog to maintain plasma glucose levels.

Many studies by Kronfeld and his colleagues have shown that dogs doing very hard work have an increased requirement for dietary protein, although there is no evidence available in the scientific literature to suggest that feeding a high protein diet will promote superior muscle development. Hard-working dogs should be fed a diet containing only a small amount of dietary fibre. A high-fibre diet will lower digestibility, lower mineral availability and increase faecal weight and the weight of the gut content. A diet containing a small amount of dietary fibre has been shown to have some benefits over diets containing no fibre in some very hard-working dogs. Table 14 shows the composition of an ideal diet for hard work and stress as suggested by Kronfeld (1982).

TABLE 14
Ideal diet for hard work and stress (Kronfeld, 1982)

Energy proportions	
Protein %	32
Fat %	51
Carbohydrate %	17
Dry matter basis	
Protein %	42
Fat %	30
Carbohydrate %	22
Fibre %	2
Ash %	4
Digestibility %	90
Main ingredients	Meat Meat by-products Grain

There is very little information available about the requirements of hard-working dogs for vitamins and minerals. If the diet is nutritionally balanced, as the dog eats to satisfy its energy requirement it will simultaneously consume all the vitamins and minerals it requires. There may be a higher iron requirement in the hard-working dog because of its involvement in haemoglobin production and oxygen transport but this has never been studied in detail, as typical working dog diets are high in meat and will therefore be high in iron. It has also been postulated that the working dog may require more vitamin E and selenium than the normal adult dog to prevent red blood cell fragility but this too has not yet been studied in detail.

Many studies in humans have shown that muscle glycogen content is an important determinant of stamina and 'carbohydrate loading' is practised by many athletes to maximize muscle glycogen before exercise. At present this dietary regimen appears to be unsuitable for dogs. An alternative strategy is to spare muscle glycogen utilization by adapting the muscle to oxidize fat. In racing dogs, fat mobilization during exhaustive exercise was found to be related to performance and to the amount of fat consumed during training.

There are many ways of making up suitable diets for working dogs. Good nutrition, by way of a balanced diet, is a fundamental starting point — a dog which is malnourished, by being underfed or overfed, will not perform to its full potential. The most important extra requirement of working dogs is for energy. Other nutrients have been poorly investigated but some may also be required in higher concentrations during periods of very hard work and stress. Foods for working dogs should be palatable, concentrated, digestible and nutritionally balanced. Such foods do not need any further supplementation. Although it is necessary to give the bulk of the food *after* working to prevent gastric dilation-volvulus and excessive defaecation, it may be a good idea for some dogs to take a small concentrated meal before working. Frequent radical changes in diet should be avoided as these may result in further digestive upsets. On days when the dogs are not being worked a smaller allowance of the same food is probably most suitable.

MINERAL AND VITAMIN SUPPLEMENTS

Contrary to many advertisements and popular beliefs, puppies and lactating bitches do not require extra large amounts of minerals and vitamins. Provided that their needs as described in Chapter 2 are met, there is no advantage to be gained by overdosing. This is not to say that owners should exclude totally the use of mineral/vitamin supplements. Where an adequate balanced diet is fed they are unlikely to be of benefit and if given in excessive amounts can do harm. But given in moderate amounts they can have a small part to play. They help satisfy the needs of some people to give what they see as extra care and can act as an insurance for those individual animals, who through some quirk, need amounts outside the usual range. Basically they are unnecessary additions to prepared foods. However, they are essential to achieving a balanced diet from fresh foods.

CONCLUSIONS

Feeding dogs and cats at various stages of their lives using prepared foods is really quite straightforward. Most pet owners are untrained in nutrition and animal husbandry and require confidence that the feed regimen which they adopt will be nutritionally sound as well as satisfying their other expectations of palatability, convenience, price and suitability for their particular domestic arrangements. The range of types and varieties of prepared foods, if made by reputable manufacturers, are nutritionally balanced (i.e. provide adequate amounts of vitamins, minerals and major nutrients in relation to their energy content) such that they can provide this convenience and reassurance. The feeding guides provided by reputable manufacturers and the amounts suggested in Tables 12 and 13 give a good starting point from which to estimate the amounts needed by any particular dog or cat. Careful observation of the animal's appearance and behaviour will enable the owner to identify the correct level of feeding for his animal, its likes and dislikes and to arrive at a satisfactory feeding regimen.

BIBLIOGRAPHY

Adkins, T. O. and Kronfeld, D. S. (1982) Diet of racing sled dogs affects erythrocyte depression by stress. *Can. Vet. J.* **23,** 260–263.

Baines, F. M. (1981) Milk substitutes and the hand rearing of orphaned puppies and kittens. *J. Small Anim. Pract.* **22,** 555–578.

Björck, G. (1984) Care and feeding of the puppy in the postnatal and weaning period. In *Nutrition and Behaviour in Dogs and Cats,* pp. 25–33. Editor R. S. Anderson. Pergamon Press, Oxford.

Blaza, S. E. and Loveridge, G. G. (1984) Feeding and care of kittens. In *Nutrition and Behaviour in Dogs and Cats,* pp. 35–40. Editor R. S. Anderson. Pergamon Press, Oxford.

Brody, S. (1945) *Bioenergetics and Growth.* Reinhold, New York.

Downey, R. F., Kronfeld, D. S. and Banta, C. A. (1980) Diet of Beagles affects stamina. *J. Am. Anim. Hospital Assoc.* **16,** 273–277.

Durrer, J. L. and Hannon, J. P. (1962) Seasonal variations in caloric intake of dogs living in an arctic environment. *Am. J. Physiol.* **202,** 375–378.

Grøndalen, J. and Hedhammar, Å. (1984) Nutrition of the rapidly grouping dog with special reference to skeletal disease. In *Nutrition and Behaviour in Dogs and Cats,* pp. 81–88. Editor R. S. Anderson. Pergamon Press, Oxford.

Hammel, E. P., Kronfeld, D. S., Ganjam, V. K. and Dunlap, H. L. (1977) Metabolic responses to exhaustive exercise in racing sledge dogs fed diets containing medium, low and zero carbohydrate. *Am. J. Clin. Nutr.* **30,** 409–418.

Kirk, R. W. (1968) In *Canine Medicine,* pp. 809–813. Editor E. J. Catcott. A.V.P., Illinois.

Kronfeld, D. S. (1973) Diet and the performance of racing sledge dogs. *J. Am. Vet. Med. Assoc.* **106,** 470–473.

Kronfeld, D. S. (1982) Feeding dogs for hard work and stress. In *Dog and Cat Nutrition,* pp. 61–73. Editor A. T. B. Edney. Pergamon Press, Oxford.

Kronfeld, D. S. and Downey, R. L. (1981) Nutritional strategies for stamina in dogs and horses. *Proc. Nutr. Soc. Australia* **6,** 21–29.

Kronfeld, D. S. and Dunlap, H. L. (1984) Common clinical and nutritional problems in racing sled dogs. In *Nutrition and Behaviour in Dogs and Cats,* pp. 89–96. Editor R. S. Anderson, Pergamon Press, Oxford.

Kronfeld, D. S., Hammel, E. P., Ramberg, C. F. and Dunlap, H. L. (1977) Haematological and metabolic responses to training in racing sledge dogs fed diets containing medium, low or zero carbohydrate. *Am. J. Clin. Nutr.* **30,** 419–430.

Loveridge, G. G. (1986) Bodyweight changes and energy intake of cats during gestation and lactation. *Anim. Technol.* **37,** 7–15.

Loveridge, G. G. (1987) Some factors affecting kitten growth. *Anim. Technol.* **38,** 9–18.

Loveridge, G. G. and Rivers, J. P. W. (1988) Energy costs of reproduction in the cat. Waltham Symposium No. 7 1985. Cambridge University Press, Cambridge.

Mapletoft, R. J., Schutte, A. P., Coubrough, R. I. and Kuhne, R. J. (1974) The perinatal period of dogs, nutrition and management in the hand rearing of puppies. *J. South African Vet. Assoc.* **45,** 183.

Orr, N. T. M. (1966) The feeding of sledge dogs on Antarctic expeditions. *Br. J. Nutr.* **20,** 1–11.

Paul, P. and Issekutz, B. (1967) Role of extramuscular energy sources in the metabolism of the exercising dog. *J. Appl. Physiol.* **22,** 615–622.

Sheffy, B. E. (1978) Symposium on Canine Paediatrics, pp. 7–29. *Nutrition and Nutritional Disorders.* Veterinary Clinics of North America 8, No.1.

Teare, J. A., Hedhammar, Å. and Krook, L. (1984) Influence growth intensity on the skeletal development in dogs: effects ascorbic acid supplementation. In *Nutrition of the Dog and Cat,* pp. 129–143. Editor R. S. Anderson. Pergamon Press, Oxford.

Thonney, M. L. (1985) Quoted in Chapter 2, Nutrient requirements and signs of deficiency. In *Nutrient Requirements of Dogs,* pp. 2–4. National Academy Press, Washington.

Thonney, M. L., Touchberry, R. W., Goodrich, R. D. and Meiske J. C. (1976) Intraspecies relationship between fasting heat production and bodyweight: a re-evaluation of $W^{0.75}$. *J. Anim. Sci.* **43,** 690.

Widdowson, E. M. (1974) Feeding the newborn: comparative problems in man and animals. *Proc. Nutr. Soc.* **33,** 275–284.

Widdowson, E. M. (1974) Food growth and development in the suckling period. In *Canine and Feline Nutritonal Requirements,* pp. 9–17. Editor O. Graham-Jones, pub. BSAVA.

CHAPTER 6

Clinical Small Animal Nutrition

PETER J. MARKWELL

INTRODUCTION

There are two main areas in which nutrition and disease can be seen to interact. The first of these is a group of conditions that arise from errors in nutrition, which fall into two categories

1. conditions associated with undernutrition, e.g. thiamin deficiency;
2. conditions associated with overnutrition, e.g. obesity.

Deficiency conditions associated with individual nutrients have been studied extensively under experimental conditions, but single deficiencies are rare and few are of practical importance clinically. One of the reasons for this is the ready availability of nutritionally complete and balanced prepared pet foods, which have greatly simplified the feeding of dogs and cats, particularly during times of high nutritional demand.

The second broad area of interaction between nutrition and disease could be termed nutritional therapeutics or special dietetics. This classification groups together conditions that arise independently of errors in the diet, but in which dietary modifications form an important part of clinical management. Examples of this group of conditions would include chronic polyuric renal failure and exocrine pancreatic insufficiency. Considerable scientific interest is currently being expressed in this field which represents an area of growing importance in veterinary medicine.

CONDITIONS RESULTING FROM NUTRITIONAL ERRORS

Obesity

Obesity has been defined as a pathological condition characterized by an accumulation of fat much in excess of that necessary for optimal body function (Mayer, 1973). Implicit in this definition is the concept of obesity as a condition detrimental to an individual's health. This factor, coupled with the regularity with which obesity is observed (Table 15), makes it the most important form of malnutrition in small animal practice.

TABLE 15
Results of surveys of the incidence of obesity in dogs

Observed Incidence	Source of data
28%	Mason 1970
34%	Anderson 1973
44%	Steininger 1981
24.3%*	Edney and Smith 1986

*Includes dogs categorized as obese or gross.

Certain subpopulations of dogs seem particularly prone to obesity, an association linked with less discriminate eating and poorer satiety control. The condition becomes more common in the older age groups, probably related to a reduction in energy expenditure through exercise. It is also much commoner in neutered animals, with a spayed bitch twice as likely to be overweight as an entire one. This may also apply to castrated males, but few data are available. Certain breeds also show a greater tendency towards obesity, including Labradors, Cairns, Cocker Spaniels and Long Haired Dachshunds (Edney and Smith, 1986).

Objective assessment of obesity in dogs presents a number of problems. Methods based on comparison with normal body weights are difficult to apply because of lack of baseline data, and other methods used in man are also inappropriate (Anderson, 1973). Use of ultrasound to measure fat thickness appears promising (Houpt and Hintz, 1978), but requires expensive equipment and has not yet been fully studied. Assessment of subcutaneous tissue overlying the ribcage, although subjective, has been used successfully (Mason, 1970) and Edney and Smith (1986), proposed a number of criteria based on observation and palpation, which were used in their extensive survey.

Obesity is actually caused by energy intake exceeding requirements at some stage of the animal's life. The excess energy is then deposited as fat. Subsequently, energy intake may not appear particularly excessive. Body weight may be stable and as the animal is likely to be fairly inactive and has a thick insulating layer of fat, energy requirements may be quite low.

The majority of cases of obesity are related to simple overfeeding, often coupled with lack of exercise. A small percentage of cases may be associated with endocrine abnormalities, e.g. hypothyroidism or insulinoma, but Armstrong and others (1951) considered that the overweight in at least 95% of human cases was accounted for by overeating at some time and it is unlikely that the proportion is any less in dogs and cats. In general, no particular feeding regime has been linked with obesity, although Mason (1970) found that obesity was more common in dogs fed table scraps and home-prepared foods than those fed proprietary canned meat. Presumably the lack of accurate

TABLE 16
Recommended energy allowances at various target weights

kg	kcal per day
4	203
8	374
12	535
16	688
20	838
24	983
28	1126
32	1267
36	1405
40	1541
44	1676
48	1810
52	1942
56	2073
60	2203

feeding guides and great variability in constituents for home prepared diets makes errors more likely.

The management of canine obesity represents an important challenge to the small animal practitioner, because the condition has been associated with a number of serious clinical problems. These include increased susceptibility to infectious disease (Newberne, 1966); articulo-locomotor problems; circulatory problems; and increased surgical risk. A link with many other clinical problems has been suggested, but has yet to be clearly demonstrated.

Once endocrine causes have been ruled out, a choice exists between two basic methods for the management of canine obesity, starvation or controlled calorie reduction.

A system for management by controlled calorie reduction was described by Edney (1974). It is divided into a number of steps.

1. Counsel the dog owner to ensure the co-operation of all people involved in feeding the dog.
2. Weigh the dog and set an initial, realistic target weight. This should be no more than 15% less than the dog's current weight. If necessary the programme can be repeated to achieve a normal bodyweight for the breed and size of dog.
3. Prescribe a diet providing 60% of the calculated maintenance requirement at the target weight (Table 16).
4. Weigh the dog weekly, at the same time of day and using the same scale on each occasion. Every time there is no loss of body weight, reduce the food allowance by a further 20%.

Results from a series of 50 dogs showed a predictable weight loss over a 3 month period using this system (Fig. 17). Any diet may be used, but practical experience has shown that a complete change of diet is more likely to be

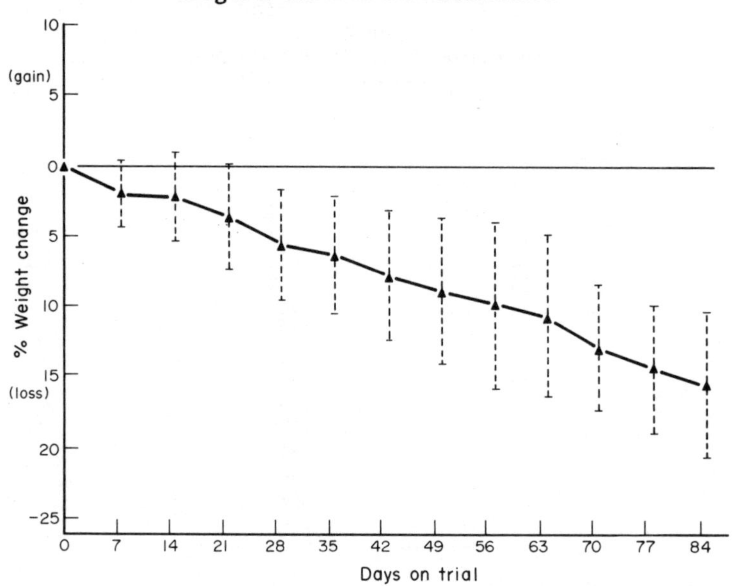

FIG. 17. Weight change during canine obesity trial. Mean and standard deviation of data from 50 clinical cases treated over a period of 3 months (Edney, Personal Communication).

successful than modifications to an existing one. The use of a diet with a high moisture content rather than a dry diet allows a greater volume of food to be eaten for any given calorie intake and thus helps to reduce problems associated with 'hunger misery' and improve owner co-operation.

Finally, the vitamin and mineral content of the diet should be considered. Most diets are balanced to provide adequate vitamins and minerals at normal calorie intakes and therefore supplementation may be required at the low calorie intakes recommended. Some foods specifically designed for use in weight reduction programmes are formulated with increased levels of vitamins and minerals to overcome this problem.

Starvation represents an alternative method to controlled calorie reduction for promoting weight loss in dogs. Hospitalization and very careful observation of the dog are necessary. All food is withheld during the period of starvation, only water and vitamin and mineral supplements are provided. Weight losses of 23% have been reported by the end of the 5th week of starvation (Lewis, 1978). No refeeding problems or adverse effects on health have been reported and biochemical studies suggest that the regime is safe (de Bruijne, 1979). However, a greater proportion of weight loss will occur from lean body mass than with controlled calorie reduction, and many owners are likely to consider starvation an unacceptable procedure on humanitarian grounds.

Fewer data exist concerning obesity and associated problems in cats, but the condition appears much less common. The reasons for this are not clear, but it

TABLE 17
Typical calcium content of some foods

Food	Calcium mg/400 kcal*
Lean beef	22.8
Dressed Tripe	500.0
Ox Liver	14.7
Milk	738.5
Egg	141.5

*Modified from Paul and Southgate (1985) with the permission of the Controller of Her Majesty's Stationery Office.

is suggested that cats have a much better ability to regulate their own energy intake, matching this more closely to requirements than dogs. One interesting link with disease that does appear to exist is an epidemiological association between obesity and an increased risk of the Feline Urological Syndrome in cats (Walker *et al.,* 1977).

Management of obesity in cats follows the same principles as in dogs. A target weight can be set and food provided giving 50%–60% of the expected calorie requirement at the target weight. An additional factor to be considered is the need to confine that cat to the house, or to hospitalize it to prevent hunting.

Nutrition and Bone Disease

A number of important skeletal diseases are associated with incorrect feeding practices. As approximately two-thirds of the skeleton is composed of inorganic material, largely salts of calcium and phosphorus, a diet deficient in these minerals can have a very marked effect on skeletal health. This is particularly apparent during periods of rapid growth in young animals. Vitamins A and D are essential for the handling and absorption of calcium and provision of a diet containing incorrect levels of these vitamins can also give rise to skeletal disease. Thus particular attention should be paid to vitamin and mineral levels in diets intended for growing dogs and cats, but it should be remembered that supplementation of an already balanced diet is not only unnecessary, even for giant breeds of dogs, but may well be detrimental.

Nutritional secondary hyperparathyroidism

Nutritional secondary hyperparathyroidism is the condition that is seen most often in both cats and dogs, despite the ease with which it can be prevented. The most common cause is the feeding of a calcium deficient diet to puppies and kittens during the period of rapid growth. Classically, this is a diet composed predominantly of fresh meat, which is a very poor source of calcium (Table 17). Frequently, there are additional nutritional complications such as the provision of excess vitamin D. Inadequate calcium intake results in a transient hypocalcaemia which stimulates release of parathyroid hormone. This acts with vitamin D to restore blood calcium towards normal. Actions include stimulation of calcium and phosphorus absorption from the gut, if they are present in the lumen; decreased calcium and increased phosphorus excretion by the kidney; and increased resorption and decreased formation of bone. The major pathological effects are seen on bone, where rapid turnover occurs with a greater acceleration of resorption compared with formation (Bennett, 1976). Clinical signs include lameness, resulting from joint and bone pain, bone deformity and in some cases pathological fractures of long bones or vertebrae. Radiography is essential for diagnosis and the major finding is loss of bone density. The growth plates appear normal, often with changes of radio-density in the adjacent metaphysis (Campbell and Griffiths, 1984).

The most important aspect of clinical management is correction of dietary imbalances by removal of excess supplements and provision of a nutritionally balanced diet which provides for the growing animal's needs. In severely affected animals, calcium may be supplemented in this diet to achieve a calcium/phosphorus ratio of 2:1 during the healing phase (Capen and Martin, 1983), but this should revert to the normal ratio of 1.2 to 1.4:1 when healing is completed. Severely affected animals should also be confined during healing to minimize the risk of pathological fractures. Analgesics may be given, but recovery from pain is usually rapid after dietary correction. Prevention of the condition is far preferable to treatment, and this may be accomplished by feeding nutritionally balanced diets to puppies and kittens. The use of good quality commercial puppy and kitten foods is recommended as providing the most convenient and simplest means of achieving this.

Rickets (osteomalacia)

Rickets is a rare disease in modern veterinary practice and uncertainty still exists over its aetiology. The major abnormality present is failure of normal mineralization of osteoid and in the young animal of the cartilaginous matrix in the epiphyseal growth plates (Campbell and Griffiths, 1984). The relative importance of calcium, phosphorus and vitamin D is not clear, but the condition appears to arise because of a combination of low vitamin D and calcium intake (Campbell, 1979). Rickets is difficult to distinguish clinically from nutritional secondary hyperparathyroidism and differential diagnosis requires radiography. Bone density is reduced and there may be bending of the shafts of long bones, but the major changes are seen in the epiphyseal plates which appear widened and irregular (Campbell and Griffiths, 1984). Treatment for the condition is based on provision of a nutritionally adequate diet containing the correct amounts of vitamin D, calcium and phosphorus. Once again the condition can be avoided by feeding growing animals correctly.

Hypervitaminosis A

Excessive intake of several individual nutrients is associated with disease and probably no group presents a more serious risk than the fat soluble vitamins. Unlike the water soluble vitamins, where excesses can readily be excreted, fat soluble vitamins are stored in the body and prolonged excessive intake can result in signs of toxicity.

In the case of vitamin A, excessive intake may result from overzealous supplementation of a diet, or it could result from poor formulation of a diet using excessive quantities of foods which are rich in vitamin A. Hypervitaminosis A is seen mostly in cats, although it has been recorded experimentally in dogs (Cho *et al.*, 1975). In cats it is frequently associated with feeding diets containing large quantities of liver, although there appears to be considerable individual variation in the quantity of vitamin A that must be ingested before signs of toxicity become apparent.

Clinical signs are related to the effects of vitamin A on bone growth and remodelling. Boney exostoses develop along the muscular insertions of cervical vertebrae and the long bones of the forelimbs. In kittens shortening of long bones and epiphyseal damage has also been reported in experimental studies (Clark *et al.*, 1970). Clinical signs include anorexia, weight loss, lameness, neck stiffness and cats develop an unkempt appearance, with moist eczema and alopecia (Hayes, 1978).

Treatment by dietary correction or removal of supplements may bring about amelioration of clinical signs, but severe skeletal changes may be irreversible. Actually changing the diet may prove quite difficult, because liver is highly palatable for cats. They may be very reluctant to consume alternative foods and have to be gradually 'weaned off' the liver.

Deficiency of Water Soluble Vitamins

Clinical signs associated with deficiency of one or more of the water-soluble vitamins may occasionally be observed in cats and dogs. Signs are generally non-specific and a diagnosis is often impossible based on clinical criteria alone. Anorexia, accompanied by a general lethargy is seen associated with the initial stages of many of these deficiencies, and will soon complicate the clinical picture. If signs are present, careful examination should be made of the diet. Clinical signs usually arise within a few days of the start of feeding a deficient diet because these vitamins are not stored in the body. Single deficiencies involving water soluble vitamins are quite rare, with the possible exception of thiamin in kittens, so a diet adequate in all essential nutrients should be fed where vitamin deficiencies are suspected.

Thiamin

Deficiencies of thiamin may be encountered occasionally in cats. Thiamin is more heat labile than most other nutrients, and is progressively, but not instantly, destroyed by cooking. Most manufactured foods are supplemented to compensate for possible losses, but other foods should have thiamin added after cooking. Deficiency can also arise if some types of fish, which contain the enzyme thiaminase, are fed raw. Thiaminase can be readily inactivated by heating, so feeding cooked fish is unlikely to cause problems.

Clinical signs of thiamin deficiency in cats appear within 1 to 2 weeks of the start of feeding a deficient diet. The first signs are anorexia and associated weight loss. These are followed by the development of neurological signs, including ataxia and short convulsive seizures. Untreated cases progress rapidly to the irreversible phase of semi-coma, opisthotonus and maintained extensor tone (Jubb *et al.,* 1956). Treatment is by injection of 500–1000 mμg of thiamin, which leads to rapid reversal of clinical signs unless brain damage has occurred (Edney, 1985). Dietary deficiencies should also be corrected immediately to avoid recurrence of the problem.

Vitamin E Deficiency

Pansteatitis (or 'yellow-fat disease') is an inflammatory condition of subcutaneous fat associated with feeding diets containing high levels of polyunsaturated fatty acids. It has been suggested that high levels of unsaturated fatty acids coupled with low levels of the antioxidant vitamin E, lead to deposition of ceroid pigment in adipose tissue with accompanying fat cell necrosis and inflammation (Gaskell *et al.,* 1975). Clinical signs observed in the condition include pyrexia, with accompanying anorexia or lethargy, and pain on abdominal palpation or movement. Biopsy of the subcutaneous fat reveals yellow colouration and histologically, fat cell necrosis is visible (Gaskell *et al.*, 1975). The condition has been recorded clinically in cats fed diets composed predominantly or entirely of fish, most commonly tuna, or when dietary fat is rancid (Edney, 1985). Treatment consists of removing fish from the diet and giving Vitamin E by tablet or injection. Resolution can take several weeks in some cases.

CONDITIONS BENEFITING FROM SPECIAL DIETARY MANAGEMENT

Chronic Polyuric Renal Failure

Chronic renal failure is a common syndrome in both cats and dogs. It represents the end stage of many progressive renal diseases, and occurs when some 75% of renal mass has been destroyed (Bovee, 1977).

As renal tissue is progressively destroyed, surviving nephrons undergo hypertrophy and hyperplasia, but eventually these compensatory responses are overcome, resulting in decreases in glomerular filtration rate and tubular transport and the development of azotaemia. Many substances change in concentration in azotaemia, and although it is not clear which substance or combination of substances are most toxic, it is generally agreed that many of the clinical signs are associated with the accumulation of protein metabolites.

Chronic renal failure in dogs

The major principle of dietary management in chronic renal failure is to minimize the build-up of protein catabolites by restricting dietary protein intake to the animal's requirement and providing energy from non-protein sources, thus preventing mobilization of the animal's own tissues. There are two other benefits that have been claimed for low protein diets, assistance in preventing phosphorus accumulation (discussed below), and altering the rate of progression of disease. This latter claim is based on work in other species. Studies in the dog with surgical models of renal failure have not supported this claim, with no progression of renal disease being recorded (Finco *et al.,* 1985).

Various recommendations have been made about the amount of protein that should be offered, from 0.66 g protein per kg body weight per day (Bovee, 1977), to 2.0–2.2 g protein per kg body weight per day (Polzin and Osborne, 1983). However, feeding a diet supplying 1.6 g protein per kg body weight per day over a 40 week period produced signs of protein malnutrition in dogs with surgical models of renal failure, whereas no signs were observed with diets providing 2.0 g protein per kg per day (Polzin *et al.,* 1983; Polzin and Osborne, 1986). Thus excessive restriction may be detrimental in some cases and diets providing 2.0 to 2.2 g of protein per kg per day are recommended as a starting point. These figures are only a guide and protein intake may need to be adjusted to suit the individual case, attempting to provide a balance between alleviation of clinical signs and avoidance of signs of protein malnutrition.

Protein restriction is not justified in animals in the early stages of disease, where phosphorus and urea nitrogen levels remain within normal limits, but therapy should be instituted when these values are exceeded. All protein supplied to the animal should be of high biological value, for example egg or lean meat, and poorer quality proteins should be avoided. Provision of an adequate supply of energy from non-protein sources (fat and carbohydrate) is essential to avoid mobilization of the animal's body tissues, or use of dietary protein to provide energy. A reasonable proportion of this energy should come from fat, not only is this much more energy dense that carbohydrate, reducing the volume of food that the animal needs to eat, but it also helps to make the diet more palatable. A table of calorie requirements for dogs of different weights is given in Chapter 2 but adjustments may have to be made to account for the activity, body weight and condition of the animal.

Renal secondary hyperparathyroidism develops as falling glomerular filtration rate leads to phosphorus retention and hyperphosphataemia. This results in a reciprocal decrease in blood calcium levels and stimulation of parathyroid hormone release. Parathyroid hormone acts on bone, kidney and gut to restore blood calcium levels towards normal, particularly causing resorption of bones. To control renal secondary hyperparathyroidism, restriction of phosphorus intake is necessary. This will be aided by use of a low-protein diet, but in addition to this it is frequently necessary to restrict the phosphorus content of the diet. In some cases oral phosphorus-binding agents may also be necessary, e.g. aluminium hydroxide gel. When serum phosphorus has been normalized, calcium supplementation may be commenced. Calcium may be given as calcium carbonate, at a dose of 100 mg per kg body weight per day (Bovee *et al.,* 1987). The use of preparations containing phosphorus, e.g. bone flour, should be avoided. Calcium supplementation should not commence prior to normalization of serum phosphorus because of the danger of soft tissue calcification. Vitamin D may also be given, but only when serum calcium and phosphorus levels can be monitored.

Sodium supplementation has been recommended, primarily to compensate for obligatory sodium loss occurring through damaged nephrons. In addition to this, decreased sodium intake may lead to a reduction in the ability to reabsorb bicarbonate (Schmidt and Gavellas, 1977), which could contribute to a metabolic acidosis.

Some reports have, however, suggested that the incidence of hypertension in dogs with renal conditions may be high (Weiser *et al.,* 1977; Anderson and Fisher, 1968) and although this remains very controversial, restriction of sodium intake would form an important aspect of management of this condition. Additionally, dogs with varying degrees of renal failure are able to maintain sodium balance within the normal range of dietary intake, but may not adjust rapidly to sudden changes in dietary sodium content (Cowgill, 1983). On the basis of this information it is recommended that diets are prepared and fed without addition or specific restriction of sodium unless there are clear indications to the contrary. It is also very important to avoid sudden changes in dietary sodium content, because the animal may not be able to adapt rapidly enough. Finally, supplementation with water soluble vitamins is advisable because of possible losses resulting from polyuria.

Chronic renal failure in cats

In theory, dietary manipulation based on the principles described above is also appropriate in the cat with chronic renal failure. Under practical circumstances this may prove very difficult. Cats are fastidious feeders and as low protein diets tend to have poor palatability for them, compromises may have to be made to get the cat to eat any food at all. Some suggestions for encouraging animals which are ill to eat are given in the final section of this chapter.

No data are available concerning the protein requirements of nephritic cats, but a recommended starting point is between 3.3–3.5 g of protein per kg body weight per day (Polzin and Osborne, 1986). This value should then be adjusted to provide the maximum intake that the cat will tolerate at the given level of renal failure. The other principles discussed concerning energy, vitamins and minerals may also be applied to the cat.

Diabetes Mellitus

Diabetes mellitus is quite a common endocrine disorder and it has been suggested that it affects approximately 1 in 200 dogs and 1 in 800 cats (Chastain and Nichols, 1984). The major clinical signs of diabetes mellitus are polydipsia, polyuria and weight loss. The underlying condition is one of carbohydrate intolerance due to inadequate insulin production or insulin resistance, giving rise to hyperglycaemia and glycosuria. Management of the condition generally involves daily insulin injections and careful dietary regulation.

The objective of dietary regulation is to provide a consistent calorie intake, balanced to the animal's requirement. By keeping these constant, the dosage of insulin can be adjusted to control hyperglycaemia. Little research has been done on the best diet for use in dogs and cats with diabetes mellitus, but in general prepared foods offer clear advantages over home-prepared diets, including consistency and predictability of calorie content, enabling greater accuracy in feeding.

Carbohydrate does not increase the need for insulin (unless in the form of simple sugars where hyperglycaemic peaks may be produced), provided that it doesn't increase the total energy intake (Leibetseder, 1982). Furthermore, clinical experience with diabetic dogs suggests that some are unable to maintain weight on a diet where few calories are derived from carbohydrate (Blaxter, personal cummunication). On this basis it is quite reasonable to use carbohydrate as part of the diet, and a diet of equal volumes of a canned food and mixer, (about 3:1 by weight) is recommended. This provides about 40–45% of calories from carbohydrate. A complete canned food containing cereal would also be suitable, an example of which provides approximately 50% of calories from carbohydrate. These values are of a similar order to recommendations for human diabetics that 50% or more of energy intake should come from carbohydrate (Chase, 1977).

When calculating the quantity of food to be offered to the dog, account should be taken of obesity or weight loss and the dog fed to a reasonable target weight. In addition to controlling energy input it is important to control output as far as possible, providing a constant amount of exercise each day. The benefits of including fibre in diets for human diabetics have received considerable attention. Jenkins *et al.,* (1976), demonstrated that post-prandial blood glucose responses were significantly reduced in patients consuming meals with added dietary fibre, compared with consumption of the same meals without fibre. Further work is now required with dogs and cats before specific recommendations can be made regarding fibre levels in diets for canine and feline diabetics.

In addition to providing a consistent level of feeding, timing of meals is also important. About 1/4 of the daily food allowance should be given immediately prior to the insulin injection, so that if the animal refuses to eat no injection has been given. This timing may also help to get the animal to accept the injection. The remaining food can be given later in the day when the insulin will exert its maximum effect (Bush, 1984).

The Feline Urological Syndrome

Dietary measures form an important part of the management of cats that have had the Feline Urological Syndrome (FUS), and a number of recommendations may be made to reduce the risk of recurrence. Ideally a diet should be directed towards achieving the following:

1. A high urine volume: a combination of high fluid intake and low faecal water loss will help to achieve this. A moderately high level of sodium chloride (around 3.0% of dry matter) will help to stimulate drinking and foods with a high water content e.g. canned food or fresh meat, should be provided to ensure that water is taken in with the food. Faecal water loss can be minimized by feeding a highly digestible diet to reduce faecal volume. Production of a high volume of dilute urine reduces the risk of crystal formation in the urine.
2. A low magnesium intake: magnesium is the most important mineral in relation to FUS. Studies have shown that very high concentrations of magnesium of 0.3% or more on dry matter increase the risk of FUS (Lewis and Morris, 1984) because of increased formation of struvite (magnesium ammonium phosphate) in the urine. These levels are in fact considerably in excess of the levels in reputable commercial cat foods.
3. A low urine pH: struvite is more soluble in slightly acid conditions, thus a slightly acid urine is less likely to result in precipitation of struvite crystals and thus reduces the risk of FUS recurring. Use of urinary acidifiers may be required, e.g. ammonium chloride or *DL* methionine, to achieve this. If acidifiers are used they should be given with the food and care should be taken to avoid acidosis.

In addition to the dietary measures listed above, consideration should also be given to aspects of the cat's lifestyle, such as access to the outside to allow plenty of opportunities for urination, and where possible a cat flap should be provided for this. It is also worth considering whether the cat is overweight, as this is linked epidemiologically with FUS.

Congestive Heart Failure

Dietary management can form a useful adjunct to primary therapy in congestive heart failure. Animals with congestive heart failure tend to retain

TABLE 18
Typical sodium content of some foods

Food	Sodium mg/400 kcal*
Meat	
Beef (lean)	198.4
Lamb (lean)	217.3
Chicken	267.8
Veal	403.7
Vegetables	
Beans	171.4
Cabbage	127.3
Potatoes	32.2
Cereals	
Macaroni (boiled)	27.4
Spaghetti (boiled)	6.8
Rice (boiled)	6.5
Dairy foods	
Butter	470.3
Cream (single)	79.2
Egg yolks	59.0

*Modified from Paul and Southgate (1985) with the permission of the Controller of Her Majesty's Stationery Office.

sodium and water. The mechanisms involved in this are complex, but involve activation of the renin–angiotensin system and alterations in glomerular filtration, which combine to cause sodium and water retention by the kidney (Thomas, 1977). These mechanisms ultimately lead to congestion and oedema. Restriction of dietary sodium intake has been shown to be effective in stopping sodium retention in dogs with congestive heart failure (Pensinger, 1964) and therefore it should help to prevent the development of congestion.

Therefore, the primary goal of dietary therapy for congestive heart failure is restriction of sodium intake. The degree of sodium restriction required may vary with the severity of disease, Ettinger and Suter (1970) recommend that sodium intake for a small dog should not exceed 6 mg/lb (13 mg/kg), with slightly less being given to large and giant breeds. This is equivalent to 60–70 mg/400 kcal. Table 18 lists a number of foods that could be used in diets together with their sodium contents. Actually formulating a palatable low sodium diet can be quite difficult, and one method that may be used to enhance palatability is to warm the food to 37–38°C.

In addition to the sodium content of the diet, consideration should also be given to its vitamin content. Increased levels of B group vitamins should be supplied to compensate for possible losses resulting from diuretic therapy.

The energy content of the diet is also an important consideration, and the food allowance should be carefully calculated to avoid or correct obesity or cachexia. Finally the feeding pattern should be adjusted to provide several small meals per day. The feeding of large quantities of food at any one time should be avoided, as pressure on the diaphragm from a full stomach may hamper cardiac function mechanically (Leibetseder, 1982). Sodium restriction should be used as an adjunct to diuretic therapy; it is unlikely to be rewarding when used alone.

Gastric dilation-volvulus

Gastric dilation-volvulus is a life-threatening situation and constitutes a real emergency. Most cases occur in large, deep-chested breeds such as Bloodhounds, Borzois and Great Danes. German Shepherd Dogs, Irish Wolfhounds and St Bernards are also very prone to the condition. The cause of the condition is unknown and remains the subject of some controversy. Factors that have been suggested as predisposing to the condition include postprandial exercise and excitement, overdrinking, certain types of diet, and previous gastric trauma.

The role of diet is unclear. Dry foods have been implicated but a false association may have been made because foods of this type frequently form a major part of the diet of the breeds most prone to the condition. The source of the gas present in the stomach has also been the subject of debate, but analytical studies suggest that it is derived from swallowed air and not from bacterial fermentation (Caywood *et al.*, 1977). Preliminary evidence now exists that abnormal electrical activity in gastric smooth muscle may be the underlying abnormality (Burrows, 1986).

Irrespective of cause, a number of suggestions may be made to minimize the risk of gastric dilation-volvulus in dogs.

1. Avoid all unnecessary excitement at feeding time.
2. Feed the daily allowance in two, three or even more separate meals.
3. Keep exercise periods as far away from feeding times as possible.
4. Give food to the dog so that it eats with its head up in an elevated position.
5. Feed wet foods.

Pearson (1975) recommended pre-soaking any biscuit fed to susceptible dogs.

Dietary Allergy

The most common manifestations of dietary allergy are skin changes and gastrointestinal signs. Clinical signs associated with the skin include pruritis, erythema and papules (White, 1986), but the picture may frequently be complicated by self-inflicted injury. Gastrointestinal signs include vomiting and diarrhoea which may occur within a few hours of exposure to the allergen (Walton, 1967).

Diagnosis of dietary allergy requires identification of the allergen by feeding an exclusion diet and challenging with suspect foods. This process requires very strict dietary management and a sustained effort on behalf of both owner and veterinary surgeon. The animal is placed on a test diet for up to 2 weeks and monitored closely for remission of clinical signs. Lamb and rice has traditionally been used as a basis for this diet, but other single protein sources to which the animal is unlikely to have been exposed would be satisfactory. It is essential that the chosen diet forms the sole source of nutrients so cats will have to be confined to the home or hospitalized for the duration of the test. Secondary bacterial infections and skin changes should be treated at the same time. Failure to improve may reflect the fact that the chosen diet contains the allergen, or that the diagnosis is incorrect. If improvements are noted, potential allergens may be introduced singly into the diet allowing several days for changes to appear. It is logical to start with the most common allergens, milk (casein), beef and wheat gluten. Permanent exclusion of all identified allergens from the diet is essential because only small quantities are necessary to trigger a response.

Multiple sensitivities appear uncommon in dogs and cats, although Walton (1967) did identify 2 dogs from a series of 82 that were allergic to both milk and beef.

Gastro-Intestinal Disease

Dietary modifications can form an important part of investigation or management of diarrhoea. Diarrhoea is a clinical sign resulting from an increase in the frequency of defaecation and in the fluidity of faeces. Normal faeces typically contain 70% water and only a relatively small rise to about 85% is necessary to produce diarrhoea. Thus, diarrhoea can result from any factor which influences fluid absorption or secretion in the bowel. This can arise from many causes including bacteria, viruses, parasites, toxins, dietary factors, neoplasia and a variety of specific conditions such as exocrine pancreatic insufficiency and lactase deficiency.

Withholding food has traditionally formed a part of the management of diarrhoea, with only water or an electrolyte solution being provided. This is followed by the introduction of a highly digestible diet containing good quality protein and with a low fat level. A general recommendation can also be made to avoid lactose-containing products and ensure a low level of crude fibre (Leibetseder, 1982), and diets containing liver should also be avoided as some individuals do not tolerate this food. Examples of suitable foods would be eggs, lean meat, cooked rice (which is usually well digested). Some commercial foods with a low fat content may also be appropriate. Food should be provided on a 'little and often' basis, the daily intake being spread over 3–4 months.

In addition to these general comments dietary therapy is appropriate in a number of specific diseases.

Lactose intolerance

Disaccharides such as lactose have to be hydrolyzed to monosaccharides before they can be absorbed from the intestinal lumen. The enzymes responsible for this are located at the epithelial brush border. Some individual dogs and cats have reduced quantities of the enzyme lactase, and therefore have a limited ability to hydrolyze lactose. Undigested lactose remains within the intestinal lumen allowing proliferation of lactose-fermenting bacteria and acting as an osmotic laxative, producing a wet, profuse diarrhoea. Treatment is straightforward, remove the source of lactose, remembering that it is contained in a variety of dairy products and not just in milk.

Exocrine pancreatic insufficiency

Exocrine pancreatic insufficiency is diagnosed more frequently in dogs than in cats. In dogs three distinct entities will give rise to clinical signs of exocrine pancreatic insufficiency; pancreatic hypoplasia; pancreatic degenerative atrophy; and chronic pancreatitis (Murdoch, 1979). Pancreatic degenerative atrophy is particularly common in the German Shepherd Dog, in which breed it is inherited as an autosomal recessive condition. The clinical signs that arise in each case are similar and are associated with a decrease in the secretion of a number of digestive enzymes which leads to poor digestion of food. The poorly digested food remains in the bowel lumen, retaining fluid osmotically. Steatorrhoea is also present due to poor fat digestion and animals therefore tend to produce large volumes of pale, evil smelling, fatty faeces. Poor digestion of food coupled with vitamin losses in the faeces result in malnutrition, and animals with exocrine pancreatic insufficiency are usually underweight and have poor coats.

The basis of therapy is replacement of digestive enzymes at the time of feeding using one of a number of oral preparations. Dietary measures consist of providing a readily digestible diet with a low fat content. Increased levels of vitamins, particularly of the fat soluble group should also be provided, and up to double the normal recommended allowance may be given.

Feeding Animals Which Are Ill

One of the problems frequently associated with feeding animals which are ill is inappetence. This is particularly the case with cats which are notoriously fastidious feeders. This problem may be exacerbated when attempts are being made to feed home-prepared low sodium or low protein diets, neither of which are particularly palatable even to healthy dogs and cats.

With the exception of some of the conditions already noted the requirements of sick animals are generally higher than those of healthy ones. This is partly associated with the stress accompanying disease and partly with factors such as the increase in metabolic rate that accompanies fever. For example, the general level of metabolism is increased by about 10% for every 1°C rise in body temperature (Leibetseder, 1982).

A number of basic principles can be applied to encourage animals to eat, and to ensure as far as possible that increased nutrient demands are being met.

1. Particular attention should be paid to the diet composition to ensure that it is complete and balanced. Where appetite is depressed additional vitamins and minerals may be given if there is no clinical contraindication.
2. The diet should be of high digestibility.
3. The diet should have a high nutrient density, therefore minimizing the amount that the animal has to eat to obtain necessary nutrients.
4. Feed on a 'little and often' basis, dividing the total daily intake into 3–4 meals.
5. Warm food to 38–39°C if necessary, but not beyond this temperature. Temperature has a marked effect on food preference and this can help considerably with tempting inappetent animals to eat.
6. Include animal fat in the diet if there is no contraindication to it. In addition to being a rich source of energy, it helps to increase palatability.
7. Feed foods with a high moisture content. They tend to be more palatable than dry foods.
8. Remove food that is not eaten after 10–15 minutes. Fresh food offered later is likely to prove more acceptable.

In some cases despite all these measures animals may remain inappetent, and syringe or tube feeding may become necessary.

BIBLIOGRAPHY

Anderson, L. J. and Fisher, E. W. (1968) The blood pressure in canine interstitial nephritis. *Res. Vet. Sci.* **9,** 304–313.

Anderson, R. S. (1973) Obesity in the dog and cat. *Veterinary Annual,* 14th Edition, p. 182. J. Wright, Bristol.

Armstrong, D. B., Dublin, L. I., Wheatley, G. H. and Marks, H. H. (1951) Obesity and its relation to health and disease. *J. Am. Med. Assoc.* **147,** 1007–1014.

Bennett, D. (1976) Nutrition and bone disease in the dog and cat. Vet. Record **98,** 313–320.

Bovee, K. C. (1977) Diet and Kidney Failure In *The Kal Kan Symposium for the Treatment of Dog and Cat Disease,* pp. 25–28. Kal Kan Foods Inc., Vernon, CA 90058, U.S.A.

Bovee, K. C., Bower, C. R. and Bower, H. (1987) Dietary management of renal failure. *Tijdschrift voor Diergeneeskunde* **112,** 825–915.

de Bruijne, J. J. (1979) Biochemical observations during total starvation in dogs. *Int. J. Obesity* **3,** 239–247.

Burrows, C. F. (1986) Diseases of the canine stomach. In *The Veterinary Annual,* 26th Edition, pp. 270–282. J. Wright, Bristol.

Bush, B. M. (1984) Endocrine system. In *Canine Medicine and Therapeutics,* Editors Chandler, E. A., Sutton, J. B. and Thompson, D. J., 2nd Edition, p. 206. Blackwell, Oxford.

Campbell, J. R. (1979) Undermineralization in dogs and cats. In *Proceedings of the Second*

Waltham Symposium. p. 16. Waltham Centre for Pet Nutrition, Waltham-on-the-Wolds, Freeby Lane, Melton Mowbray, Leicestershire.

Campbell, J. R. and Griffiths, I. R. (1984) Bones and muscles. In *Canine Medicine and Therapeutics,* Editors Chandler, E. A., Sutton, J. B. and Thompson, D. J., 2nd Edition, p. 138. Blackwell, Oxford.

Capen, C. L. and Martin, S. L. (1983) Calcium regulating hormones and diseases of the parathyroid gland. In *Textbook of Veterinary Internal Medicine,* Editor Ettinger, S. J., p. 1550. W. B. Saunders, Philadelphia.

Caywood, D., Teague, H. D., Jackson, D. A., Levitt, M. D. and Bond, J. H., Jr (1977) Gastric gas analysis in the canine gastric dilation-volvulus syndrome. *J. Am. Anim. Hospital Assoc.* **13,** 459.

Chase, H. P. (1979) Diabetes and diet. *Food Technol.* **33,** 60–64.

Chastain, C. B. and Nichols, L. E. (1984) Current concepts on the control of diabetes mellitus. *Vet. Clin. N. Am.* **14,** 859–872.

Cho, D. Y., Frey, R. A., Guffy, M. M. and Leipold, H. W. (1975) Hypervitaminosis A in the dog. *Am. J. Vet. Res.* **36,** 1597–1603.

Clark, L., Seawright, A. A. and Gartner, R. J. W. (1970) Longbone abnormalities in kittens following vitamin A administration. *J. Comp. Pathol.* **80,** 113–121.

Cowgill, L. D. (1983) Diseases of the kidney. In *Textbook of Veterinary Internal Medicine,* Editor Ettinger, S. J., p. 1793. W. B. Saunders, Philadelphia.

Edney, A. T. B. (1974) Management of obesity in the dog. *Vet. Med./Small Anim. Clinician* **49,** 46–49.

Edney, A. T. B. (1985) Feline nutrition and disease. In *Feline Medicine and Therapeutics,* Editors Chandler, E. A., Gaskell, C. J. and Hilbery, A. D. R., p. 339. Blackwell, Oxford.

Edney, A. T. B. and Smith, P. M. (1986) Study of obesity in dogs visiting veterinary practices in the United Kingdom. *Vet. Rec.* **118,** 391–396.

Ettinger, S. J. and Suter, P. F. (1970) Low sodium diets and other drugs and methods indicated in cardiac therapy. In *Canine Cardiology,* p. 257. W. B. Saunders, Philadelphia.

Finco, D. R., Crowell, W. A. and Barsanti, J. A. (1984) Effect of three diets on dogs with induced chronic renal failure. *Am. J. Vet. Res.* **46,** 646–653.

Gaskell, C. J., Leedale, H. L. and Douglas, S. W. (1975) Pansteatitis in the cat: a report of four cases. *J. Small Anim. Pract.* **16,** 117–121.

Hayes, K. C. (1978) Nutritional problems in cats: taurine deficiency and vitamin A excess. *Can. Vet. J.* **23,** 2–5.

Houpt, K. A. and Hintz, H. F. (1978) Obesity in dogs. *Canine Pract.* **5,** 54–58.

Jenkins, D. J. A., Leeds, A. R., Gassull, M. A., Wolever, T. M. S., Goff, D. V., Alberti, K. G. M. M. and Hockaday, T. D. R. (1976) Unabsorbable carbohydrate and diabetes: decreased post-prandial hyperglycaemia. *Lancet* **ii,** 172–174.

Jubb, K. B., Saunders, L. Z. and Coates, H. V. (1956) Thiamin deficiency encephalopathy in cats. *J. Comp. Pathol.* **66,** 217–227.

Leibetseder, J. L. (1982) Feeding animals which are ill. In *Dog and Cat Nutrition,* Editor Edney, A. T. B. 1st Edition, p. 85. Pergamon, Oxford.

Lewis, L. D. (1978) Obesity in the dog. *J. Am. Anim. Hospital Assoc.* **14,** 402–409.

Lewis, L. D. and Morris, M. L. (1984) Feline urological syndrome: causes and clinical management. *Vet. Med.* **79,** 323–337.

Mason, E. (1970) Obesity in pet dogs. *Vet. Rec.* **86,** 612–616.

Mayer, J. (1973) Obesity. In *Modern Nutrition in Health and Disease,* Editors Goodhart, R. S. and Shils, M. E., p. 625. Lea and Febiger, Philadelphia.

Murdoch, D. B. (1979) Alimentary tract and associated glands. In *Canine Medicine and Therapeutics,* Editors Chandler, E. A., Sutton, J. B. and Thompson, D. J., 1st Edition, p. 285. Blackwell, Oxford.

Newberne, P. M. (1966) Overnutrition and resistance of dogs to distemper virus. *Fed. Proc. Fedn. Am. Soc. exp. Biol.* **25,** 1701–1710.

Paul, A. A. and Southgate, D. A. T. (1985) *McCance and Widdowson's The Composition of Foods.* H.M.S.O., London.

Pearson, H. (1975) Gastric dilation and torsion. *Pedigree Digest* **2,** 6.

Pensinger, R. R. (1964) Dietary control of sodium intake in spontaneous congestive heart failure in dogs. *Vet. Med./Small Anim. Clinician* **59,** 752–784.

Polzin, D. J. and Osborne, C. A. (1983) Conservative medical management of canine chronic polyuric renal failure. In *Current Veterinary Therapy VIII,* Editor Kirk, R. W., p. 997. W. B. Saunders, Philadepĥia.
Polzin, D. J., Osborne, C. A., Stevens, J. B. and Hayden, D. W. (1983) Influence of modified protein diets on the nutritional status of dogs with induced chronic renal failure. *Am. J. Vet. Res.* **44,** 1694–1702.
Polzin, D. J. and Osborne, C. A. (1986) Update — conservative medical management of chronic renal failure. In *Current Veterinary Therapy XI,* Editor Kirk, R. W. p. 1167. W. B. Saunders, Philadephia.
Schmidt, R. W. and Gavellas, G. (1977) Bicarbonate reabsorption in experimental renal disease: effects of proportional reduction of sodium or phosphate intake. *Kidney Int.* **12,** 393–402.
Steininger, E. (1981) Die Adipositas und ihre Diatetische Behandlung. *Wiener Tierärztlicher Monatschrift* **68,** 122–130.
Thomas, W. P. (1977) Low sodium diets. In *Current Veterinary Therapy VI,* Editor Kirk, R. W., p. 342. W. B. Saunders, Philadelphia.
Walker, A. D., Weaver, A. D., Anderson, R. S., Crighton, G. W., Fennell, C., Gaskell, C. J. and Wilkinson, G. T. (1977) An epidemiological survey of the feline urological syndrome. *J. Small Anim. Pract.* **18,** 283–301.
Walton, G. S. (1967) Skin responses in the dog and cat to ingested allergens. Observations on one hundred confirmed cases. *Vet. Rec.* **81,** 709–713.
Weiser, M. G., Spangler, W. L. and Gribble, D. H. (1977) Blood pressure measurement in the dog. *J. Am. Vet. Med. Assoc.* **171,** 364–368.
White, S. D. (1986) Food hypersensitivity in 30 dogs. *J. Am. Vet. Med. Assoc.* **188,** 695–698.

APPENDIX I

Legislation

At EEC and national levels, the basic legislation affecting the pet food industry is that which relates to farm animals and covers the manufacture and marketing of its products. This agrees with the industry's own view but it is important also that the EEC Commission and national governments take into account the special characteristics and requirements of the prepared pet food industry within the directives for feed for farm animals. (Legislation for example should, and to some extent already does, take into account the standards established for the nutrition of pet animals, that prepared pet foods are purchased alongside and from the same distribution outlets as human foods, that health/safety regulations affecting food-producing animals are not relevant to pet animals.)

Because only part of the basic legislation is specific to prepared pet foods, some countries have chosen to prepare codes of practice. An example of this are the codes of practice and product descriptive terms agreed by the United Kingdom Pet Food Manufacturers' Association (PFMA).

Legislation at EEC and national levels is concerned with the health and safety of products, the use of additives, marketing practices and safety/health regulations governing the importation of raw materials. At EEC level the basic legislation affecting the industry is contained in the following four Directives.

1. **Additives** – which is concerned with the use of additives in all feeds including prepared pet foods.
   November 23, 1970 (70/524/EEC).
   This directive has already been subject to amendments on 43 occasions. The main amendments are contained in 2 annexes. Annex 1 being a permanent list, annex 2 being a temporary list.
2. **Undesirable substances** – concerned with the fixing of maximum permitted levels for undesirable substances and products in feedingstuffs. December 17, 1973 (74/63/EEC).
3. **Marketing of straight feedingstuffs** – concerned with labelling and the distribution of straight feedstuffs. November 23, 1976 (79/101/EEC).

4. **Marketing of compound feedingstuffs** – concerned with labelling and the distribution of compound feedingstuffs. April 2, 1979 (79/373/EEC).

Other directives which also affect the evaluation of prepared pet foods cover sampling and analytical methods: packaging materials: safety and health regulation relating to meat importations and consumer protection aspects common to all industries (which includes directive 79/112/EEC concerned with the labelling and advertising of foods).

In the USA, Official Pet Food Regulations have been prepared and approved by AAFCO (American Association of Feed Control Officials) in conjunction with industry's representatives under the auspices of the American Petfood Institute (PFI) and are currently in official status under a new Uniform Feed Bill approved by AAFCO (1987). Comprehensive regulations exist governing definitions and terms (PF1), label format and labelling (PF2), brand and product names (PF3), expression of guarantees (PF4), ingredients (PF5), directions for use (PF6) and drugs and pet food additives (PF7).

In addition protocols for adequate testing of pet food products have been developed by a committee of nutritionists from the PFI in the USA at the request of AAFCO. This committee was concerned only with the testing of foods for normal healthy animals. Thus the protocols presented cover gestation/lactation, growth and maintenance of adult animals. It was considered that protocols for testing products intended for the dietary management of disease states lay outside the directive given to the committee.

The AAFCO protocols or variants thereof have tended to be used by reputable pet food manufacturers in many other countries to assess nutritional adequacy. However it is important to appreciate that AAFCO protocols are the minimum necessary to substantiate particular product nutritional claims and more complex testing is frequently undertaken by some manufacturers in order to ensure life-long health.

APPENDIX II

Definitions

The AAFCO definition of a complete food is 'a nutritionally adequate feed for animals other than man; by specific formula is compounded to be fed as the sole ration and is capable of maintaining life and/or for promoting production without any additional substance being consumed except water'. The EEC Council Directive on marketing of compound feedingstuffs (1979) define a 'complete feedingstuff' as 'a compound feedingstuff which by reason of its composition is sufficient to ensure a daily ration'. A 'daily ration' means 'the total quantity of feedingstuff, expressed on a 12% moisture basis, required by an animal of a given kind, age group and level of production in order to satisfy its average daily nutritional needs'.

Therefore both the AAFCO (1987) and EEC Feedingstuffs Regulations (1979) definitions of complete feedstuffs mean the same thing. Some dog foods are not intended to form the whole diets for dogs and cats and are termed complementary foods, examples of these could be low protein dog biscuits or some meat-based canned foods. The United Kingdom Feedingstuffs Regulations (1986) defines 'complementary feedingstuff' as a compound feedingstuff which, by reason of its composition, is not sufficient to ensure a daily ration unless it is used in combination with other feedingstuffs.

The term 'balanced' is defined by AAFCO (1987) as a term that may be applied to a diet, ration or feed having all known required nutrients in proper amount and proportion based upon recommendations of recognized authorities in the field of animal nutrition, such as the National Research Council (NRC), for a given set of physiological animal requirements. The species for which it is intended and the functions such as maintenance or maintenance plus production (growth, foetus, fat, milk, eggs, wool, feathers or work) shall be specified. In the case of dog foods the most authoritative collection of published data on nutritional requirements is the NRC (1985) document, *Nutrient Requirements for Dogs.* This document is widely used by formulators of dog foods to establish balanced nutrient profiles in products. A sister document (NRC 1986) and equally useful reference has been published for cats, entitled the *Nutrient Requirements of Cats.*

APPENDIX III

Biological Trial Procedures in Dog and Cat Nutrition

A. DIGESTIBILITY TRIALS

The objective of digestibility trials is to measure the availability of the nutrient content of a food product, complete diet or raw material (i.e. food ingredient) to the animal. The nutrient content of the test food is determined by analysis of representative samples. Measured amounts are individually fed to six dogs or cats for a minimum period of 14 days (dogs) or 21 days (cats). Faeces are quantitatively collected for the last 7 days (dogs) or 14 days (cats) of each feeding period and analysed for their nutrient content. The average daily amount of nutrient apparently absorbed into the animal's body is calculated from the difference between nutrient intake as food and nutrient output as faeces. This amount expressed as a proportion of intake is the digestibility percentage or availability percentage of the food as fed. Where the test product or raw material is not suitable to be fed as the only food, because for example, of serious nutritional imbalance or because it is unpalatable, then a series of trials is necessary in which the test material is fed in different proportions with a basal diet. It is then possible to calculate the digestibility and the digestible nutrient content by 'difference' or by 'regression' methods.

Since faeces do not consist only of undigested, unabsorbed material but contain cell debris and material excreted into the digestive tract, the difference between intake and output measured in this way is defined as apparent digestibility or apparent absorption. To measure true absorption or true digestibility, it is necessary to use control diets free of the nutrient being studied to establish the size of output when intake is zero. For most practical purposes apparent digestibility is the measurement used as it measures the net amount of digestion. Within a species, digestibility values are largely independent of the individual animal and are more a characteristic of the food.

Metabolism or balance trials are extensions of digestibility trials in that urine output is collected as well as faeces. It is then possible to draw up a balance sheet showing intake of nutrients from the food or test material and total output in faeces and urine. The difference between intake and output is regarded as the amount of nutrient retained in the body. (There are losses other than faecal and urinary losses which have to be accounted for. These are mainly via the integument, such as skin and hair.)

In young growing animals, the balance trial permits such estimates as the amounts of protein or mineral retention and provides a good method of estimating the protein quality of foods or ingredients. The technique is less valuable with adults which are not normally depositing or accumulating protein as a result of growth. The errors then become considerable. Metabolism trials also permit the measurement of energy lost in urine and of water turnover, both of which are useful in comparing foods of different kinds such as fresh foods, canned or dry products.

Although a knowledge of food digestibility does not give a complete picture of its nutritional value, it allows digestive and absorptive efficiency to be measured. Taken together with detailed analyses of micronutrient content this information gives a reliable indication of whether or not foods or combinations of foods are providing adequate nutrition for adult maintenance and evidence of their likely suitability for supporting growth or lactation.

Additionally they indicate any adverse or beneficial effects on the amount and consistency of faeces produced. Since each digestibility trial takes a maximum of 21 days, a large number of raw materials and types of food can be assessed.

Digestibility Trials in Dogs

Dogs are fed once daily at 9.00 a.m. They are penned individually in concrete floored pens, equipped with an automatic water drinker and electrically heated bed. Faeces are collected first thing in the morning and then at regular intervals throughout the day, often within minutes of being voided. Faeces voided outside the working day are collected the next morning. Faecal collections are stored in a refrigerator to minimize spoilage.

At the end of the 7 day collection,the faeces are mixed and representative samples obtained for freeze drying and analysis. Freeze dried samples are used for all chemical analysis and for energy determination by adiabatic bomb calorimetry. Food samples are prepared in the same way.

Digestibility Trials in Cats

Essentially the same procedure is followed for cats. Cats are penned individually. They are provided with a separate small tray in which they have been trained to defaecate and urinate.

The amounts of food given are maintained constant for the whole of the trial period so that by the time faeces are collected after a 7 day run-in period, a day's faeces can be related to a day's food. The amount fed is calculated to provide enough energy for adult maintenance (80 kcal/kg bodyweight).

B. GROWTH TRIALS

Biological trials are necessary because it is possible that foods or combination of foods, which on analysis would provide enough nutrients to meet the estimated requirements of an animal, might prove to be unsuitable when fed over a long period. They may not be of high enough palatability so that animals may not be consistently willing to eat enough or they may have some unknown contamination or deficiency which affects their feeding value. An extended feeding trial with growing animals is able to determine a food's ability to support normal, healthy growth and development.

Growth Trials with Dogs

A minimum of six puppies per diet regime, and more usually nine to twelve, are used. More than one breed may be used, the commonest being Labradors and Beagles. Dogs must be pure-bred. An appropriate design which will allow statistical evaluation of the results is necessary. Randomized block designs are used in which dogs on test and control diets are matched by breed, age, initial weight, litter origin and sex whenever possible. Equal sex distribution is preferred but not essential. Trials usually being at 7–8 weeks of age. The 7th week is then used to accustom the puppies to being fed individually and to train them to eat the appropriate test or control diet prior to the start of the trial.

Feeding

No other food is given other than the appropriate control or test diet regime. Drinking water is available *ad libitum*. Although group feeding is permissible and easier to carry out, it does not allow statistical comparison of food and nutrient intakes unless there are several groups on each diet. It is therefore less useful in interpretation of results and hence puppies are usually fed on an individual basis.

Control and test animals are fed in the same way. It may be *ad libitum*, to appetite 2–4 times a day or to a scale of intake based on the weight of the puppy. The latter method is usually adopted since it prevents the dogs becoming obese if the food is extremely palatable. Other feeding regimes that meet the needs of good husbandry would be acceptable provided that both control and test groups are treated in the same manner.

No nutritional supplements are given at any time. No medication is given except under the direction of a veterinary surgeon and all such medication is recorded. Routine vaccination and anti-worm preparations are administered as normal for the animal. Trials normally continue for a minimum of 10 weeks.

Records

Individual bodyweight is recorded at the beginning of the trial and at least at weekly intervals (usually two weighings per week). Body length (nose to rump) is measured at 2 weekly intervals. The daily food consumption for each dog is recorded.

Blood characteristics, haemoglobin (Hb), packed cell volume (PCV), red and white cell count, mean corpuscular haemoglobin concentration (MCHC), mean corpuscular volume (MCV), mean cell haemoglobin (MCH) as a minimum are measured at 7, 13 and 18 weeks of age. Total plasma protein, urea, alanine transaminase and alkaline phosphatase are also measured at the same times.

Each puppy is examined by a veterinary surgeon at the beginning and end of trial. The examinations cover eyes, ears, mouth, rectal temperature, skin and coat condition, congenital/hereditary defects, circulatory, respiratory, digestive, urino-genital systems and musculo-skeletal development. A daily record of faeces consistency is maintained.

Criteria for evaluation of results

The average weight gain or rate of weight gain of test dogs should not be significantly less than that of the control group as determined by an appropriate statistical test. The control group must be within the normal range expected for its breed. Average Hb and PCV values should not be statistically significantly below those of the controls, nor outside the normal range of dogs of similar age and breed.

Growth Trials with Cats

Kitten growth trials are basically similar to puppy growth trials in objective and design. Because cats are smaller and easier to keep, larger numbers of animals are used and with better statistical control. Kittens are usually weaned at a later age and so trials do not usually start until the animals are 10 weeks old.

A minimum of 8 kittens per diet group but normally 10 or 12 are used. Equal sex distribution is preferred. Randomized block designs are used with kittens within a block matched as far as possible for age, litter origin, weight and sex. All kittens are weaned at 8 weeks and begin growth trials by 10 weeks of age. The trials run for a minimum 10 week period.

Feeding

No food is given other than the appropriate control or test diet regime. Drinking water is available *ad libitum*. As with puppies, group feeding is possible although it does not allow statistical comparison of food and nutrient intakes unless there are several groups on each diet. It is therefore less useful in the interpretation of results and hence kittens are usually fed on an individual basis. In kitten growth trials it is the normal practice to allocate both the test and control foods *ad libitum* or to appetite at several meals. No medication should be given except under veterinary direction and then comprehensive records or the reason and effect of medication should be kept.

Records

Individual bodyweight is recorded at the beginning of the trial and at least at weekly intervals (usually two weighings per week). Food consumption is recorded daily for each kitten when fed as individuals or the average per cat.

Blood characteristics, Hb, PCV, MCHC, MCV, MCH, red and white cell counts as a minimum are measured at the start and end of the trial. Total plasma protein, urea, alanine transaminase and alkaline phosphatase are also measured at the beginning and end of each trial.

Each kitten is examined by a veterinarian at the start and end of the trial. The examinations cover eyes, ears, mouth, rectal temperature, skin and coat condition, congenital/hereditary defects, circulatory, respiratory, digestive, uro-genital systems and musculo-skeletal development.

Criteria for evaluation of results

The average weight gain and growth rate of the test group should not be less than the control group by appropriate statistical analyses.

The average Hb and PCV values of the test group shall not be less than the values for the controls by appropriate statistical analysis.

None of the test group kittens shall show abnormal muscular or skeletal development, loss of body, skin or coat condition which is ascribed to the diet. None shall have any of the characteristic signs of nutritional deficiencies.

C. REPRODUCTION TRIALS

The objective is to demonstrate that the food or feeding regime will support breeding females in good health and permit normal reproduction of healthy viable young when fed throughout pregnancy and lactation.

Reproduction Trials in Dogs

As a minimum the bitches should be fed the test food or diet regime from the beginning of oestrus until the puppies are weaned at 6–8 weeks of age. Feeding for the trials usually begins a few weeks prior to oestrus and all puppies are weaned at 6 weeks. Trials rarely extend over more than one parity.

Animals

AAFCO protocols require a minimum of six pregnant females, four of which must perform satisfactorily. Dogs are pure-bred, usually Labradors or Beagles, and are between 1 and 8 years old. Whenever possible animals are used which have previously had at least one litter.

A control group is not essential since the characteristics measured can be compared with data obtained previously, but it is desirable when animals are available for the purpose.

Feeding

Each bitch is fed individually. No supplements or medication are given except under veterinary direction; all such medication is recorded. Water is freely available, and the same formulation (though not necessarily the same batch of product) is fed throughout the trial.

At 2–3 weeks of age the puppies are gradually weaned on to the same food as the bitch. Food intake is recorded for each bitch plus puppies. Weaning should be complete by 6 weeks of age.

Bitches may be fed to appetite, to a feeding scale, by packet recommendation or *ad libitum*. Bitches during pregnancy are fed to a maintenance level for the first 35 days of pregnancy and then food is increased by 10% each week until parturition. In lactation, bitches may be fed *ad libitum*, or to a scale dependent on bitch weight and the size and weight of her litter.

Detailed records of bodyweight of bitch and puppies are maintained together with records of daily food consumption and observations on health and behaviour. Blood samples are taken from the bitch on four occasions and from puppies at weaning to check haematological and biochemical parameters. A detailed physical examination is given to each bitch by a veterinary surgeon in the week after mating, the last week of pregnancy, the first week of lactation, and at weaning. Puppies are given a detailed veterinary examination at birth and weaning.

Criteria for evaluation of results

When a control group is used the average of the test group for any record (but particularly for weight changes of bitch, weight and numbers of puppies per litter and the veterinary assessment of health) should not be significantly poorer than the average for the control group.

Usually there is no control group and the following criteria apply. The American Association of Feed Control Officials (AAFCO) protocols demand that at least two-thirds of the bitches which became pregnant, with a minimum of 4, have to meet the criteria. Individual bitches should gain weight steadily during gestation. The average weight at weaning should ideally equal the average weight at mating. Individual animals may not weigh less than 85% of their weight at mating. Each bitch must rear 75% of her one day-old puppies to 6 weeks of age with a minimum number of 4 for large breeds and 3 for small and medium-sized breeds (less than 20 kg mature weight).

Any bitch with any abnormality or loss of condition ascribed to nutrient deficiency or excess is indicative of failure of the food. Haemoglobin and PCV values must not fall below the accepted normal values for animals in the unit where the test is done. These values will be approximate to 12–13% for Hb and 30–40% for PCV.

The average birth and 6-week weight of puppies should be within the normal range for their breed. Puppies must appear normal on veterinary examination with no signs of nutritional deficiency.

Reproduction Trials in Cats

The objective with cats, as for dogs, is to demonstrate the ability of a food or feeding regime to support good reproductive performance in normal healthy animals.

Animals

A minimum of 8 pregnant mature queens must be used for each diet being tested.

Feeding

Feeding is usually *ad libitum* with the appropriate food and water freely available. During the mating period and the first 6 or 7 weeks of pregnancy cats on the same food are housed together and fed as a group. Food intake is recorded as the mean per cat. Six or seven weeks after mating the queens are penned individually until they have given birth and subsequently reared their kittens to 8 weeks of age. During this period, individual records are kept of food intake for queen plus litter.

Weight records are kept for individual queens on a weekly basis throughout the feeding trials, and on the day of mating, the day after parturition and again weekly thereafter. Kittens are weighed at weekly intervals from birth.

Veterinary examinations are made at the beginning of the trial, during the first week of lactation, and 6 weeks after parturition and at weaning. Blood samples to check on haemoglobin and PCV are also obtained on these occasions. Veterinary examinations and blood sampling of kittens are done at 6 weeks of age.

Criteria for evaluation of results

All queens should gain weight steadily during gestation. Average weight of queens at 6 weeks should ideally be not less than the average at mating. Weight loss, if any, should not be significantly greater than any corresponding weight loss of a control group. Average number and weight of kittens born alive and at 6 weeks of age, should not be significantly less than a control group. Each queen should rear at least 60% of her day-old kittens unless illness, diagnosed by a veterinarian as being unconnected with diet, is responsible for greater losses. None of the queens or kittens should show signs characteristic of nutrient deficiency.

D. ADULT MAINTENANCE TRIALS

Maintenance of weight and condition of adults is nutritionally much less demanding than for growth and reproduction. Feeding trials to demonstrate nutritional adequacy or suitability of foods for this purpose usually last for 6 months. This is because a well-nourished healthy adult will have quite large reserves of many nutrients which can be used whenever the diet is inadequate and these can mask a dietary insufficiency for long periods. It is impractical to expect regular feeding trials of test products to extend for periods of a year or more which would be necessary if marginal deficiencies in some nutrients, e.g. trace metals, iron, or major deficiencies in nutrients which are stored in the body, e.g. vitamins A or D, are to be revealed.

Therefore reliance is on digestibility tests of foods which can be done regularly and which give a good measure of the availability of most nutrients. Combined with the results of detailed chemical analysis of the nutrient content of foods, which show if known safe levels are present, good assessment of suitability of adult maintenance is obtained. A supporting and relatively short feeding trial to demonstrate adequate intake is usually available as a result of the digestibility trial.

APPENDIX IV

Nutrients in dog food formulated for growth

Required Minimum Concentrations of Available Nutrients in Dog Food Formulated for Growth

Nutrient	Per 1000 kcal ME	Dry Basis (3.67 kcal ME/g)
Protein[a]		
Indispensable amino-acids		
Arginine	1.37 g	0.50%
Histidine	0.49 g	0.18%
Isoleucine	0.98 g	0.36%
Leucine	1.59 g	0.58%
Lysine	1.40 g	0.51%
Methionine-cystine	1.06 g	0.39%
Phenylalanine-tyrosine	1.95 g	0.72%
Threonine	1.27 g	0.47%
Tryptophan	0.41 g	0.15%
Valine	1.05 g	0.39%
Dispensable amino-acids	17.07 g	6.26%
Fat	13.6 g	5.0%
Linoleic acid	2.7 g	1.0%
Minerals		
Calcium	1.6 g	0.59%
Phosphorus	1.2 g	0.44%
Potassium	1.2 g	0.44%
Sodium	0.15 g	0.06%
Chloride	0.23 g	0.09%
Magnesium	0.11 g	0.04%
Iron	8.7 mg	31.9 mg/kg
Copper	0.8 mg	2.9 mg/kg
Manganese	1.4 mg	5.1 mg/kg
Zinc[b]	9.7 mg	35.6 mg/kg
Iodine	0.16 mg	0.59 mg/kg
Selenium	0.03 mg	0.11 mg/kg
Vitamins		
A	1,011 IU	3,710 IU/kg
D	110 IU	404 IU/kg
E[c]	6.1 IU	22 IU/kg
K[d]	—	—
Thiamin[e]	0.27 mg	1.0 mg/kg
Riboflavin	0.68 mg	2.5 mg/kg
Pantothenic acid	2.7 mg	9.9 mg/kg

Niacin	3 mg	11.0 mg/kg
Pyridoxine	0.3 mg	1.1 mg/kg
Folic acid	0.054 mg	0.2 mg/kg
Biotin[d]	—	—
Vitamin B_{12}	7 μg	26 μg/kg
Choline	340 mg	1.25 g/kg

[a]Quantities sufficient to supply the minimum amounts of available indispensable and dispensable amino-acids as specified above. Compounding practical foods from natural ingredients (protein digestibility ± 70%) may require quantities representing an increase of 40% or greater than the sum of the amino-acids listed above, depending upon ingredients used and processing procedures.

[b]In commercial foods with natural ingredients resulting in elevated calcium and phytate content, borderline deficiencies were reported from feeding foods with less than 90 mg zinc per kg.

[c]A fivefold increase may be required for foods of high PUFA content.

[d]Dogs have a metabolic requirement, but a dietary requirement was not demonstrated when foods from natural ingredients were fed.

[e]Averages must be considered to cover losses in processing and storage.

APPENDIX V

Minimum Requirements for Growing Kittens[a]

Nutrient	Unit	Amount
Fat[b]		
Linoleic acid	g	5
Arachidonic acid	mg	200
Protein[c] (N × 6.25)	g	240
Arginine	g	10
Histidine	g	3
Isoleucine	g	5
Leucine	g	12
Lysine	g	8
Methionine plus cystine (total sulphur amino-acids)	g	7.5
Methionine	g	4
Phenylalanine plus tyrosine	g	8.5
Phenylalanine	g	4
Taurine	mg	400
Threonine	g	7
Tryptophan	g	1.5
Valine	g	6
Minerals		
Calcium	g	8
Phosphorus	g	6
Magnesium	mg	400
Potassium[d]	g	4
Sodium	mg	500
Chloride	g	1.9
Iron	mg	80
Copper	mg	5
Iodine	μg	350
Zinc	mg	50
Manganese	mg	5
Selenium	μg	100
Vitamins		
Vitamin A (retinol)	mg	1 (3333 IU)
Vitamin D (cholecalciferol)	μg	12.5 (500 IU)
Vitamin E[e] (α-tocopherol)	mg	30 (30 IU)
Vitamin K[f] *(phylloquinone)*	μg	100
Thiamin	mg	5
Riboflavin	mg	4

Vitamin B_6 (pyridoxine)	mg	4
Niacin	mg	40
Pantothenic acid	mg	5
Folacin (folic acid[f])	μg	800
Biotin[f]	μg	70
Vitamin B_{12} (cyanocobalamin)	μg	20
Choline[g]	g	2.4
Myo-inositol[h]	—	—

[a]Based on a diet with an ME concentration of 5.0 kcal/gdry matter fed to 10- to 20-week-old-kittens. If dietary energy density is greater or lesser, it is assumed that these requirements should be increased or decreased proportionately. Nutrient requirement levels have been selected based on the most appropriate optimal response (i.e., growth, nitrogen retention, metabolite concentration or excretion, lack of abnormal clinical signs, etc.) of kittens fed a purified diet. Some of these requirements are known adequate amounts rather than minimum requirements. Since diet processing (such as extruding or retorting) may destroy or impair the availability of some nutrients, and since some nutrients, especially the trace minerals, are less available from some natural feedstuffs than from purified diets, increased amounts of these nutrients should be included to ensure that the minimum requirements are met. The minimum requirements presented in this table assume availabilities similar to those present in purified diets.

[b]No requirement for fat is known apart from the need for essential fatty acids and as a carrier of fat-soluble vitamins. Some fat normally enhances the palatability of the diet.

[c]Assuming that all the minimum essential amino-acid requirements are met.

[d]The minimum potassium requirement increases with protein intake.

[e]This minimum should be adequate for a moderate to low-fat diet. It may be expected to increase three- to four-fold with a high PUFA diet, especially when fish oil is present.

[f]These vitamins may not be required in the diet unless antimicrobial agents or antivitamin compounds are present in the diet.

[g]Choline is not essential in the diet but if this quantity of choline is not present the methionine requirement should be increased to provide the same quantity of methyl groups.

[h]A dietary requirement for myo-inositol has not been demonstrated for the cat. However, almost all published studies in which purified diets have been used have included myo-inositol at 150 to 200 mg/kg diet and no studies have tested a myo-inositol-free diet.

NOTE: The minimum requirements of all the nutrients are not known for the adult cat at maintenance. It is known that these levels of nutrients are adequate and that protein and methionine can be reduced to 140 and 3 g/kg diet, respectively. It is likely that the minimum requirements of all the other nutrients are also lower for maintenance than for the growing kitten.

The minimum requirements of all the nutrients are not known for reproduction for the adult male or female cat. It is known that with the following modifications the Nutrient Allowances as recommended in the 1978 NRC report are adequate for gestation and lactation (in units/kg purified diet, note these recommendations are based on 4.0 kcal/g dry diet): arachidonate, 200 mg: zinc, 40 mg; vitamin A, 5500 IU; and taurine, 500. It is probable that the minimum requirements for growing kittens in this table would satisfy all requirements for reproduction if the following were modified as shown: vitamin A, 6000 IU/kg diet, and taurine, 500 mg/kg diet.

APPENDIX VI

Energy requirements of dogs and cats at different physiological states

Energy requirements of dogs at different physiological states (kcal ME/day)

Body weight (kg)	Body weight (lb)	Adult maintenance	Late pregnancy	Peak lactation
1	2.2	100	150	300
2	4.4	184	276	552
3	6.6	263	394	789
4	8.8	339	508	1016
5	11.0	412	618	1237
6	13.2	484	726	1452
7	15.4	554	831	1663
8	17.6	623	935	1870
9	19.8	691	1037	2074
10	22.0	759	1138	2276
11	24.2	825	1237	2475
12	26.4	891	1336	2672
13	28.6	956	1433	2867
14	30.8	1020	1530	3060
15	33.0	1084	1626	3251
16	35.2	1147	1721	3441
17	37.4	1210	1815	3630
18	39.6	1272	1909	3817
19	41.8	1334	2002	4003
20	44.0	1396	2094	4188
21	46.2	1457	2186	4372
22	48.4	1518	2277	4555
23	50.6	1579	2368	4736
24	52.8	1639	2459	4917
25	55.0	1699	2548	5097
26	57.2	1759	2638	5276
27	59.4	1818	2727	5454
28	61.6	1877	2816	5631
29	63.8	1936	2904	5808
30	66.0	1995	2992	5984
31	68.2	2053	3080	6159
32	70.4	2111	3167	6334
33	72.6	2169	3254	6507

34	74.8	2227	3340	6681
35	77.0	2284	3427	6853
36	79.2	2342	3513	7025
37	81.4	2399	3598	7197
38	83.6	2456	3684	7368
39	85.8	2513	3769	7538
40	88.0	2569	3854	7708
42	92.4	2682	4023	8046
44	96.8	2794	4191	8382
46	101.2	2906	4358	8717
48	105.6	3016	4525	9049
50	110.0	3127	4690	9380
52	114.4	3237	4855	9710
54	118.8	3346	5019	10038
56	123.2	3455	5182	10364
58	127.6	3563	5345	10689
60	132.0	2671	5506	11013
62	136.4	3778	5668	11335
64	140.8	3885	5828	11656
66	145.2	3992	5988	11976
68	149.6	4098	6148	12295
70	154.0	4204	6306	12613
72	158.4	4310	6465	12929
74	162.8	4415	6622	13245
76	167.2	4520	6780	13559
78	171.6	4624	6936	13872
80	176.0	4728	7093	14185
82	180.4	4832	7248	14497
84	184.8	4936	7404	14808
86	189.2	5039	7559	15117
88	193.6	5142	7713	15426
90	198.0	5245	7867	15735
92	202.4	5347	8021	16042
94	206.8	5449	8174	16348
96	211.2	5551	8327	16654
98	215.6	5653	8479	16959
100	220.0	5754	8632	17263

Reproduced with the permission of the National Academy of Sciences.

Energy requirements of cats at different physiological states (kcal ME/day)

Body weight (kg)	Body weight (lbs)	Adult maintenance	Late pregnancy	Peak lactation
2.5	5.5	200	340	600
3.0	6.6	240	408	720
3.5	7.7	280	476	840
4.0	8.8	320	544	960
4.5	9.9	360	612	1080
5.0	11.0	400	680	1200

Reproduced with the permission of the National Academy of Sciences.

APPENDIX VII

Further Reading List

Anderson, R. S. (1973) Obesity in the dog and cat. *Veterinary Annual, 14,* 182–186. Wright, Bristol.

Anderson, R. S. (Editor) (1975) *Pet Animals and Society*. Pergamon, Oxford.

Anderson, R. S. (1980) Dietary aspects of diabetes in the dog. *Pedigree Digest* **7**, 5–7.

Anderson, R. S. (Editor) (1980) *Nutrition of the Dog and Cat*. Pergamon, Oxford.

Anderson, R. S. (1980) Water content in the diet of the dog. *Veterinary Annual, 21,* 171–178. Wright, Bristol.

Anderson R. S. (Editor) (1984) *Nutrition and Behaviour in Dogs and Cats*. Pergamon, Oxford.

Anderson, R. S. (1984) Nursing sick animals. *Pedigree Digest* **11** (full reference), 3–5.

BVA (1987) *Small Animal Nutrition* Suppl. *Vet. Rec.* BVA London.

Barnett, K. C. and Burger, I. H. (1980) Taurine deficiency retinopathy in the cat. *J. Small Anim. Pract.* **22,** 521–534.

Blaza, S. E. (1981) The nutrition of giant breeds of dog. *Pedigree Digest* **8**, 8–9.

Blaza, S. E. (1982) Energy requirements of dogs in cool conditions. *Canine Pract.* **9,** 10–15.

Burger, I. H. (1982) Effect of processing on the nutritive value of food: meat and meat products. In *Handbook of Nutritive Value of Processed Food: Vol. 1: Food for Human Use,* (Edited by Rechcigl, M., Jr). CRC Series in Nutrition and Food. CRC Press, Boca Raton, USA.

Burger, I. H. (1984) Home guard — domestic chemical hazards for dogs and cats. *Pedigree Digest* **11,** 13–14.

Burger, I. H. (1984) The zinc story. *Pedigree Digest* **11,** 6–7.

Burger, I. H. and Flecknell, P. A. (1985) Poisoning in the cat. In *Feline Medicine and Therapeutics,* (Edited by Chandler, E. A., Gaskell, C. J. and Hilbery, A. D. R.), pp. 321–338. Blackwell, Oxford.

Chandler, E. A., Sutton, J. B. and Thompson, D. J. (Editors) (1984) *Canine Medicine and Therapeutics,* 2nd Ed. Chapter 20, Nutrition and Disease by A. T. B. Edney, pp. 538-550. Blackwell, Oxford.

Chandler, E. A., Gaskell, C. J. and Hilbery, A. D. R. (Editors) (1985) *Feline*

Medicine and Therapeutics, Chapter 33, Nutrition and Disease by A. T. B. Edney, pp. 339–351. Blackwell, Oxford.

Edney, A. T. B. (1978) Small animal nutrition, the present state. *Veterinary Annual, 18*, 285–289. Wright, Bristol.

Edney, A. T. B. (Editor) (1979) *Diarrhoea in the Dog.* Waltham Symposium No. 1, published by Waltham Centre for Pet Nutrition.

Edney, A. T. B. (Editor) (1980) *Over- and Under-Nutrition.* Waltham Symposium No. 2., published by Waltham Centre for Pet Nutrition.

Edney, A. T. B. (1980) Rearing motherless puppies. *Pedigree Digest* **7**, 7–9.

Edney, A. T. B. (1980) (Editor) Recent advances in feline nutrition. Waltham Symposium No. 4. *J. Small Anim. Pract.* **23,** 517–613.

Edney, A. T. B. (1984) Gastric dilation, torsion. *Pedigree Digest* **11,** 8–9.

Edney. A. T. B. (Editor) (1987) Canine Development Throughout Life. Waltham Symposium No. 8 *J. Small Anim. Pract.* **28**, 947–1064.

Edney, A. T. B. and Hughes, I. B. (1986) *Pet Care.* Blackwell, Oxford.

Edney, A. T. B. and Mugford, R. A. (1987) *Practical Guide to Dog and Puppy Care.* Salamander, London.

Fiennes, R. (1981) Bothie in Antarctica. *Pedigree Digest* **8**, 7.

Gannon, J. R. (1981) Nutritional requirements of the working dog. *Veterinary Annual*, 21, 161–166. Wright, Bristol.

Grant, T. G. (1986) A behavioural study of a beagle bitch and her litter during the first 3 weeks of lactation. *Anim. Technol.* **37**, 157–167.

Heath, J. S. (Editor) (1978) *Aids to Nursing Small Animals and Birds,* 2nd Ed., Chapter on Feeding, by A. T. B. Edney, pp. 2–12. Ballière Tindall, London.

Holme, D. W. (1981) Diets for growing dogs. *Veterinary Annual,* 21, 157–160. Wright, Bristol.

Kendall, P. T. (1980) New developments in kitten nutrition and feeding. *Pedigree Digest* **7**, 10.

Kendall, P. T. (1980) Some nutritional differences between the dog and cat. *Pedigree Digest* **6**, 4–6.

Kendall, P. T. (1980) Too much supplementation can be harmful. *Pedigree Digest* **7**, 3–5.

Kendall, P. T. (1982) Dietary fibre — a role in the diet of dogs. *Pedigree Digest* **9**, 5–7.

Kendall, P. T. (1983) Influence of nutrition on coat condition in dogs and cats. *Pedigree Digest* **10**, 4–7.

Kendall, P. T. (1984) The use of fat in dog and cat diets. In *Fats in Animal Nutrition,* (Edited by J. Wiseman), pp. 383–404. Butterworth, London.

Kendall, P. T., Blaza, S. E. and Smith, P. M. (1983) Comparative digestibility energy requirements of adult Beagles and domestic cats for bodyweight maintenance. *J. Nutr.* **113,** 1946–1955.

Kendall, P. T., Blaza, S. E. and Smith, P. M. (1983) Influences of level of intake and dog size on digestibility of dog foods. *Br. Vet. J.* **139,** 361–362.

Kendall, P. T., Burger, I. H. and Smith, P. M. (1985) Methods of estimation

of the metabolizable energy content of cat foods. *Feline Pract.* **15**, 38–44.
Kendall, P. T. and Holme, D. W. (1982) Studies on the digestibility of soya bean products, cereals, cereal and plant by-products in diets of dogs. *J. Sci. Food Agric.* **33,** 813–822.
Kendall, P. T., Holme, D. W. and Smith, P. M. (1982). Methods of prediction of the digestible energy content of dog foods from gross energy value, proximate analysis and digestible nutrient content. *J. Sci. Food Agric.* **33,** 823–831.
Lane, D. R. (1985) *Jones's Animal Nursing,* 4th Ed. Chapter on Feeding by R. S. Anderson, pp. 181–204. Pergamon, Oxford.
Loveridge, G. G. (1984) The establishment of a barriered, respiratory disease-free cat breeding colony. *Anim. Technol.* **35,** 83–92.
MAFF (1976) *Manual of Nutrition,* 8th Ed. HMSO, London.
McLean, J. G. (1981) Essential fatty acids in the dog and cat. *Veterinary Annual,* 21, 167–170. Wright, Bristol.
Messent, P. R. (1981) A review of recent developments in human companion animal studies. *Proc. 5th Kal Kan Symposium,* published by Kal Kan Foods, Vernon, USA.
National Research Council (1985) *Nutrient Requirements of Dogs.* National Academy of Sciences, Washington.
National Research Council (1986) *Nutrient Requirements of Cats.* National Academy of Science, Washington.
Thorne, C. J. (1985) Cat feeding behaviour. *Pedigree Digest* **12**, 4–6.
Turner, W. T. (1980) *How to Feed Your Dog.* Popular Dogs, London.
Walker, A. D. (1980) *Fit for a Dog.* Davis-Poynter, London.
Walton, G. S. (1976) Food allergies in the dog and cat. *Pedigree Digest,* **3,** 5–10.
Watson, A. D. J. (1981) Nutritional osteodystrophies in dogs. *Veterinary Annual,* 21, 209–218. Wright, Bristol.
Wilkinson, G. T. (1981) Nutritional deficiencies in the cat. *Veterinary Annual,* 21, 183–189. Wright, Bristol.
Wilkinson, G. T. (1984) *Diseases of the Cat,* 2nd Ed. Chapter on Nutrition by P. P. Scott, p. 10. Pergamon, Oxford.
Wright, M. and Walters, S. (1980) *The Book of the Cat.* Pan Books, London.

Index